MEET THE SASQUATCH

MEET THE SASQUATCH

Christopher L. Murphy

In association with
John Green and Thomas Steenburg

ISBN 0-88839-573-6 (paperback trade edition)
ISBN 0-88839-580-9 (patron edition)
ISBN 0-88839-574-4 (limited edition)

Copyright © 2004 Christopher L. Murphy

Cataloging in Publication Data

Murphy, Christopher L.
  Meet the Sasquatch / Christopher L. Murphy ; in association
    with John Green and Thomas Steenburg.

Prepared in conjunction with a sasquatch exhibit held at the
  Vancouver Museum, British Columbia, Canada, 2004.
Includes bibliographical references and index.
ISBN 0-88839-580-9 (bound).—ISBN 0-88839-573-6 (pbk.).—
ISBN 0-88839-574-4 (limited ed.)

   1. Sasquatch. 2. Sasquatch—Pictorial works. I. Green, John, 1927–
II. Steenburg, Thomas N. (Thomas Nelson) III. Vancouver Museum
IV. Title.

QL89.2.S2M874 2004   001.944   C2004-902230-X

All rights reserved. No part of this publication may be reproduced, stored in a retrieval system or transmitted, in any form or by any means, electronic, mechanical, photocopying, recording, or otherwise, without the prior written permission of Hancock House Publishers.
Printed in China—JADE

Editor: Thomas Steenburg
Book & Cover Design: Rick Groenheyde
Front cover images: Sasquatch portrait by Peter Travers based on the creature seen in frame 352 of
            the Patterson/Gimlin Film; background footprint cast, Abbott Hill (1982),
            D. Heryford (see pages 105 and 114).
Back cover images: Pastel enhancement of the creature seen in frame 352 of the Patterson/Gimlin
            film by Christopher L. Murphy (see page 70).

*We acknowledge the financial support of the Government of Canada through the
Book Publishing Industry Development Program (BPIDP) for our publishing activities*

Vancouver Museum Disclaimer: This compendium is a private publication that has been produced in conjunction with an exhibition at the Vancouver Museum, but the opinions expressed therein are solely those of Chris Murphy, in association with John Green and Thomas Steenburg, and do not necessarily reflect those of the Vancouver Museum.

Published simultaneously in Canada and the United States by

HANCOCK HOUSE PUBLISHERS LTD.
19313 Zero Avenue, Surrey, B.C. V3S 9R9

HANCOCK HOUSE PUBLISHERS
1431 Harrison Avenue, Blaine, WA 98230-5005

(604) 538-1114   Fax (604) 538-2262
(800) 938-1114   Fax (800) 983-2262
Web Site: www.hancockhouse.com   *Email:* sales@hancockhouse.com

Contents

ABOUT THE AUTHOR AND HIS ASSOCIATES 7
ACKNOWLEDGMENTS 8
INTRODUCTION 9

CHAPTER 1 FIRST NATIONS SASQUATCH REFERENCES 10
First Nations Stone Carvings 10
First Nations Petroglyphs 13
First Nations Wood Carvings 17
First Nations Pictographs 19
Sasquatch in Northwest Coast Native Mythology 22

CHAPTER 2 EARLY WRITTEN RECORDS 24
Early Explorers and Travelers 24
Early Newspaper Reports 25

CHAPTER 3 THE SASQUATCH "CLASSICS" 28
Fred Beck and the Apemen of Mt. St. Helens 28
Albert Ostman's Incredible Journey 30
John W. Burns and the Chehalis First Nations People 31
The Ruby Creek Incident 35
The William Roe Experience 36
Jerry Crew and the Birth of the name "Bigfoot" 37

CHAPTER 4 ORGANIZED EXPEDITIONS TO FIND THE SASQUATCH 38

CHAPTER 5 THE PATTERSON/GIMLIN FILM 40
Account of the Filming Adventure 41
The First Film Screening to Scientists 49
Photographs from the Patterson/Gimlin Film 51
Aerial View of the Film Site 56
Patterson's Camera and Filming Speed 57
The Film Site Model 58
Dimensions of the Creature seen in the Patterson/Gimlin Film 65
The Creature's Walking Pattern 66
The Wood Fragment 68
Artistic Images of the Creature in the Patterson/Gimlin Film 70
Authoritative Conclusion on the Patterson/Gimlin Film 72
    Dmitri Bayanov and Igor Bourtsev (Russian Hominologists) 72
    Dr. Dmitri D. Donskoy (USSR Central Institute of Biomechanics) 74
    Dr. Donald W. Grieve (Reader in Biomechanics) 76
    J. Glickman (Forensic Examiner) 80
    Dr. Grover S. Krantz (Anthropologist) 82
Major Hoax Claims Regarding the Patterson/Gimlin Film 84

CHAPTER 6 BIGFOOT GOES DIGITAL 90

CHAPTER 7	FOR THE RECORD — SASQUATCH INSIGHTS	96
CHAPTER 8	THE PHYSICAL EVIDENCE AND ITS ANALYSIS	102
	Footprints and Casts	102
	Footprint Cast Gallery	105
	Footprint and Cast Album	110
	The Fahrenbach Findings	124
	Dr. D. Jeffrey Meldrum and the Footprint Facts	129
	Handprints	143
	The BFRO Skookum Cast	145
	Sasquatch Hair Analysis	152
	Sasquatch Beds, Nests, Bowers or Hollows	156
	Sasquatch Sounds	160
	Sasquatch Sustenance	161
	Sasquatch Speculations	164
	Sasquatch Sighting and Track Reports	167
	Sasquatch and the Smithsonian Institution	172
	Sasquatch Protection	175
	Sasquatch Roots	178
CHAPTER 9	TRIBUTES — AMERICAN AND CANADIAN RESEARCHERS	180
	The Incomparable Bob Titmus	181
	John Green, The Legend Among Us	183
	The Intrepid Dr. Grover S. Krantz	188
	The Relentless René Dahinden	190
	Tom Steenburg, The Giant Hunter	198
	Daniel Perez, The Tireless Investigator	200
	Richard Noll — A Foremost Field Researcher	202
	Northern Exposure — The Dauntless J. Robert Alley ..	204
	Ray Crowe and the International Bigfoot Society	206
	Matt Moneymaker and the Bigfoot Field Researchers Organization	208
	Bobbie Short, The Lady with a Mission	209
	Paul Smith, The Artistic Visionary	211
CHAPTER 10	INTRIGUING ASSOCIATIONS	214
	The Russian Snowman	214
	Hominology in Russia	217
	The Yeti	225
CHAPTER 11	CONCLUSION	228
BIBLIOGRAPHY		231
PHOTOGRAPHS/ILLUSTRATIONS — COPYRIGHTS/SOURCES		233
INDEX		237

About the Author and his Associates

Christopher L. Murphy retired in 1994 after 36-years service with the British Columbia Telephone Company (now Telus). During his career, he authored four books on business processes and after retirement taught a night school course on vendor quality management at the British Columbia Institute of Technology. An avid philatelist, Chris has authored several books on Masonic Philately. He is president of the Masonic Stamp Club of New York.

Chris met René Dahinden in 1993 and then worked with René in producing posters from the Patterson/Gimlin film and marketing sasquatch footprint casts. In 1996, Chris re-published Roger Patterson's book, *Do Abominable Snowmen of America Really Exist?*, and Fred Beck's book, *I Fought the Apemen of Mt. St. Helens*. In 1997, Chris published *Bigfoot in Ohio: Encounters with the Grassman*, a book he authored with Joedy Cook and George Clappison of the Ohio Bigfoot Research and Study Group.

In 2000 Chris embarked on a project to assemble a comprehensive pictorial presentation on the sasquatch. This initiative led to the 2004 sasquatch exhibit at the Vancouver Museum, British Columbia and this publication.

John Green is a retired newspaperman. A graduate of both the University of British Columbia and Columbia University, he became interested in the sasquatch in 1957 while he was owner/publisher of the *Agassiz-Harrison Advance* newspaper and has spent the past 38 years as one of the most active participants in all aspects of the sasquatch investigation. His personal odyssey has ranged from hunting the sasquatch in the wild to following their trail across the continent, to tracking them by computer, and he has been the leader in the long attempt, only now beginning to bear fruit, to involve the scientific community in the search. He has written several books on the sasquatch and is considered a pre-eminent authority in the field, having been keynote speaker at all three of the major scientific symposiums so far held on this subject. John and his wife, June, live in Harrison Hot Springs, British Columbia. A profile of John's sasquatch-related activities is included in this volume.

Thomas Steenburg went directly from high school into the Princess Patricia Canadian Light Infantry, lst Battalion, Calgary, Alberta. After leaving the army in 1986, he authored three books on sasquatch sightings in both Alberta and British Columbia. He moved to Mission, British Columbia in 2002, primarily to facilitate conducting personal sasquatch research in this province. Tom has made many presentations at sasquatch symposiums and he is one of the few researchers who actively pursues evidence of the creature in its wilderness domain. A profile on Tom's activities is included in this volume.

Acknowledgments

Firstly, I wish to express my sincere thanks and gratitude to my associates John Green and Tom Steenburg, who have worked so closely with me on all aspects of this very extensive project.

My appreciation is also expressed to the numerous research contributors, especially Dmitri Bayonov, Igor Bourtsev and the late René Dahinden, Dr. Grover S. Krantz, and Bob Titmus.

And a very special thank you is extended to:

PETER TRAVERS who kindly provided our cover picture and other excellent drawings used in this work.

YVON LECLERC for his significant contributions and many excellent illustrations. Yvon lives in Quebec, Canada. He is a specialist in fossil imprints. His assistance on this project was invaluable.

Finally, special mention and thanks certainly needs to be expressed to some researchers and others for their contributions, support and encouragement:

> Joedy Cook
> Erik Dahinden
> Dr. Henner Fahrenbach
> Robert Gimlin
> Doud Hajicek
> David Hancock
> Dr. Jeffrey Meldrum
> Richard Noll
> Patricia Patterson
> Daniel Perez

Introduction

Nature guards her secrets well. Certainly, in many of the vast forest regions of North America, little sunlight, let alone the eyes of man, can penetrate anywhere near the forest floor. Nature is also a meticulous housekeeper. Everything that can be disposed of virtually disappears. The carcasses of animals are immediately consumed by other animals, leaving little, if any evidence that the bodies were there.

All wild animals, when alive, are also experts at disappearing. Some rely on camouflage, some on stealth and others are effectually transparent. In addition, they all share a natural instinct not to be seen. Man, with his technology, is the only creature that can stack the odds in his favor, but time and time again he has been outsmarted.

This work is about a creature we call the sasquatch (or bigfoot) that with nature's help appears to have us outsmarted. Thousands of people claim they have seen a sasquatch and multitudes of footprints testify to its passage. We even have photographs and sound recordings. Further, many well-qualified people have studied the findings and have declared that the creature definitely exists. Despite all of this evidence, however, governments and major research institutions remain unconvinced of the reality of the creature. There has never been a government sponsored initiative to find a sasquatch.

One thing is certain — the mystery is far beyond the possibility of a "hoaxing." Mainly, there are just too many credible sightings over too vast an area, over too many years, to even consider this possibility.

This book provides a highly illustrated account of what we know about the sasquatch from ancient times to the present. The work was prepared in conjunction with a sasquatch exhibit (provided by the authors) held at the Vancouver Museum, British Columbia, Canada during 2004.

It is sincerely hoped that the material we present will arouse more interest in the field of sasquatch studies and move us closer to resolving what is certainly North America's greatest mystery.

Note on Terminology: *The words "sasquatch" and "bigfoot" are interchangeable. Generally speaking, "sasquatch" is the Canadian name for the creature and "bigfoot" is the American name. Wherever appropriate, I have used the term "sasquatch." Further, I have chosen not to capitalize either word but have left them capitalized in quoted or reprinted material. I have also chosen to consider the word "sasquatch" and "bigfoot" as both singular and plural terms to avoid the cumberson or inappropriate terms "sasquatches," "bigfoots" or "bigfeet." However, in some cases, for general clarity and understanding, I have used "sasquatch creatures" and "bigfoot creatures" for a plural reference.*

First Nations Sasquatch References

There is no hard evidence that any recognized primates (other than human beings) have ever inhabited North America. Nevertheless, early North American First Nations people seem to have depicted what may be ape-like creatures in their art. Furthermore, through oral legends these people have passed on the tales of "wild men of the woods" for countless generations. It might be reasoned that their inspiration was brought about by sasquatch sightings.

FIRST NATIONS STONE CARVINGS

It is possible that ancient carved stone heads made by First Nations people in the Columbia River valley (Oregon and Washington) depict sasquatch creatures. Several heads have been discovered of which a photograph of one head and drawings of two are shown here. One of the heads (not shown) has been dated at between 1500 B.C. and A.D. 500. It is reasonable to assume that the dating of all of the heads would also be in this same time period.

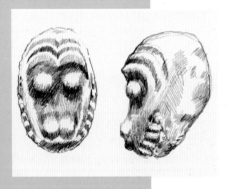

The contention that the images are just abstract or fanciful images of known animals is countered with the argument that other stone heads depict known animals that are fully recognizable. We are therefore led to the conclusion that First Nations people intentionally depicted some sort of primate other than a human being. However, there are no other known primates in North America, other than possible sasquatch creatures, therefore a connection has been reasoned. Given the estimated age of the heads, we can exclude First Nations people being aware of other primates (i.e., creatures of this nature being brought to North America by Europeans or others). However, while unlikely, we cannot exclude pet monkeys brought back from South America during early transmigration.

The first published mention of the carvings occurred in 1877 in an address[1] by Q.C. Marsh, a pioneer paleontologist. Marsh stated: *Among many stone carvings which I saw there (Columbia River) were a number of heads, which so strongly resemble those of apes that the likeness at once suggests itself.* Further, Emeritus Professor of Anthropology, Roderick Sprague (University of Idaho), states: *Several prehistoric carvings collected in the lower Columbia River valley share non-human but anthropoid features. A relationship between these stone heads and Sasquatch phenomena is suggested.*[2]

---

1 *American Association for the Advancement of Science*, Nashville, TN
2 *Manlike Monsters on Trial*, Halpin/Ames, University of B.C. Press (1980)

This haunting image created by Yvon Leclerc recalls the hairy man in the legends of First Nations people. A "wild man of the woods" is reflected in the art of native people and it is now seen that the sasquatch was probably the source of the images. Many present First Nations people are firm believers in the creature's existence and hold it sacred in their beliefs.

OFFICIAL MUSEUM DESCRIPTION OF THE STONE FOOT

NUMBER: QAD 92

AREA OR TRIBE: Lillooet

ARTIFACT: Ceremonial bowl. Medicine man's ceremonial stone.

SIZE: L: 22.4cm

W:17.8cm

Th/H: 6.5cm

DESCRIPTION: Bottom surface concave, shaped like a man's foot, with four toes; large toe broken off; heel of foot also broken off. Upper surface decorated with an oblong flower design, the center of which is an oblong concavity.

DATE COLL: 1947

COLLECTOR: Mr. S. H. Gibbs

DONOR: Mr. S. H. Gibbs

REPLICAS: Princeton Museum, May/78

We also have this remarkable stone foot that shows some resemblance to a sasquatch footprint, although it appears too short for a normal print. Here, however, it might be reasoned that it was patterned after a print made with the foot bent at the mid-tarsel break.

Stone Foot
Left, view from under foot; right, view from above

All that is known of the artifact is that it came from Lillooet, British Columbia. It was given to the Vancouver Museum in 1947. John Green provides the following analysis:

It may be just the product of the artist's imagination, but seen from the bottom it seems to be a skillful representation of a natural object, with none of the stylized effects of the top view. Such a foot could not be the foot of any known animal, as the base of the toes that is broken off is far larger than that of any of the other toes. Plainly it was a "big toe." A human foot is not a good match either. The stone foot has a heavy pad of flesh under the toes for two thirds of their length, with a deep crease dividing the pad from the ball of the foot, and its toenails (which do not show well in the pictures) are on the rear half of the terminal section of the toes, starting almost at the joints, instead of being right at the front.

This fragment can be matched fairly closely with a standard sasquatch print, but to do so requires that the missing part at the back be almost as long as what remains. To me the pattern at the top suggests that not very much is missing, and so does the shallow bowl in the bottom of the foot, which is complete in this fragment. There is more damage to the fragment than the two large missing pieces, so I could not tell whether one side at the back is already starting to round off short. From the bottom it looks as if it might be, but from the top it does not. Favoring the possibility that the original carving was much longer is the very fact that it is broken. It would be more likely for a long object to break in the middle than for a short one to break near the end.

In a letter to the Vancouver Museum dated October 11, 1972, Dr. Grover S. Krantz, Washington State University, Department of Anthropology states:

The appearance of the underside of this foot resembles the footprints of the legendary sasquatch, and this may be the earliest known record of man's concern over footprints of this type.

FIRST NATIONS PETROGLYPHS

First Nations petroglyphs are images etched in stone. They depict numerous designs and representations of animals and people. Some images appear to represent sasquatch-like creatures. As the images are thousands of years old we must again wonder at the source of the imagery.

Somewhat similar to features seen on the stone heads is a petroglyph carving in Bella Coola, British Columbia. The carvings in the area show many images of human faces with what might be termed "normal" human features. This carving, however, is distinctly different. The carvings are not attributed to the local Bella Coola First Nations people.

This petroglyph is said to represent the "Hairy Man." It is located at Painted Rock which is on the Tule River First Nations Reservation, California (Sierra Nevada foothills). Painted Rock is the site of many First Nations pictographs which are presented in the next section.

New Mexico Petroglyphs

The following material was provided by Robert W. Morgan, co-founder and current president of the American Anthropological Research Foundation Inc., a Florida not-for-profit foundation formed in 1975.[3] Robert experienced his first sasquatch sighting in 1957 and began active research in 1968. He directed four of the American Yeti Expeditions and continues active field research.

Along the Rio Grand River north of the Pueblo Indian village of Cochiti, New Mexico, is a valley now flooded by one of the world's largest earthen dams. This is the only area wherein lies the ancient legend of Gashpeta, the place where an old, starving and possibly crippled giant hairy "cannibal" woman had once terrorized the Cochiti tribe. Having been left behind or somehow separated from her own group of those whom we now call the sasquatch, she began raiding the nearby pueblo to steal unwary children and food from the elderly.

One version of this legend relates that desperate shamans of the tribe called upon the fabled "Twins" to help save the villagers. With the aid of the villagers, these Twins managed to shut the raiding cannibal giantess up in her cave. Initially, one shaman sacrificed himself by going inside to put her to sleep. As she slept, they built a great fire near the cave entrance. Before she awakened they swung three huge boulders down to seal her inside with the bones of her victims.

If this legend is true, the cave may hold the bones of a giant desert sasquatch. The cave was reportedly flooded in the early 1980s. All of the photographs that follow were taken in 1972.

"As she slept, they built a great fire near the cave entrance. Before she awakened they swung three huge boulders down to seal her inside with the bones of her victims."

3 See www.trueseekers.org and www.aarf-usa.org

The whistling mouth symbol seen here is identical to the Canadian representations of the dreaded D'sonoqua, the "Cannibal Giant" of the Kwakiutl Tribe, yet these petroglyphs were found in New Mexico, USA, over 2,000 miles/3,200km south.

Two petroglyph footprints. The print on the left appears to be human. It contrasts dramatically with the print on the right which shows some similarity to the stone foot previously discussed.

The fat belly of a satisfied giant.

These stone plugs mark what is called the entrance to the giantess' tomb. When Morgan asked permission to remove them to see if bones existed, his request was denied by certain elders who feared that the spirit of the evil giantess would be released. Unfortunately, the dam raised the river waters to the extent that any bones and even the petroglyphs may now be lost forever.

The "power stone" set in place across the site that magically seals the feared cannibal giantess in her cave.

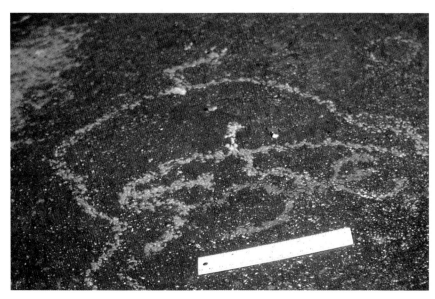

The image seen here represents the capture of a smaller human by the cannibal giantess.

Transmigration Route

If sasquatch do exist in North America, they have been here a very long time and would predate all of the foregoing stone carvings. We can reason that the creatures came here from Eurasia by crossing the land bridge that connected Eurasia to what is now Alaska (the present Bering Strait). This passage was usable for at least 20,000 years and indeed was used both ways by human hunters on the trail of arctic game. However, by about 8000 B.C. anyone or anything that was in North America was here to stay if they did not possess a boat. By this time, melting ice sheets had drastically raised the sea level so the Bering Strait area could no longer be crossed on foot.

Given the 8000 B.C. "no return" time frame, we can say that sasquatch have been here for at least 10,000 years. As to the maximum time, it is probably around 30,000 years, given the land bridge existed for about 20,000 years.

This image created by Pete Travers might give us a possible insight into the appearance of North America's earliest primates. Did some of them get "locked in time" and are now the elusive sasquatch?

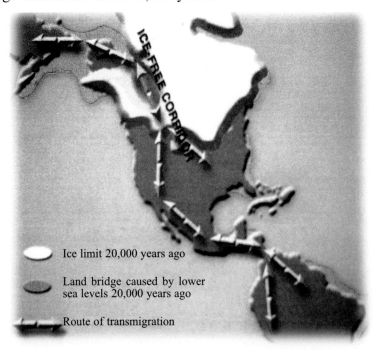

Ice limit 20,000 years ago

Land bridge caused by lower sea levels 20,000 years ago

Route of transmigration

SOME WILD CREATURE WANDERING OUR FORESTS DEFINITELY CAME FROM ASIA

JOHN GREEN TELLS US: *I was present when droppings of any unusual appearance were collected in northern California and shipped off for examination. The report that came back was that the material was the remains of fresh water plants, and that it contained eggs of parasites otherwise known only from some North American tribal groups in the northwestern U.S., pigs from south China, and pigs and people from southwest China.*

Since the material did not appear to have originated with either people or pigs, and the Asian parasites that laid the eggs did not get here by themselves, we have to wonder if the sasquatch is a highly likely suspect for the droppings.

FIRST NATIONS WOOD CARVINGS

Images of ape-like creatures have also found their way onto early wood carvings of First Nations people in British Columbia.

The most intriguing woodcarving is this Tsimshian mask discovered in British Columbia in the early part of the last century. Other than sasquatch, the only plausible explanation for the source of the image is a pet monkey brought to North America by an early European sailor.

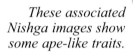

These associated Nishga images show some ape-like traits.

This sasquatch mask was created by Ambrose Point of the Chehalis First Nations people in 1938. It is quite large, undoubtedly reflecting the size of the creature it represents.

Seen here on the left is a Kwakiutl First Nations dance mask. It represents the "buck'was" or Wild Man of the Woods. On the right is a Kwakiutl heraldic pole that shows D'sonoqua (the cannibal woman). D'sonoqua is the main crest of the Nimpkish First Nations people.

This Delaware First Nations carving is the top of a 10-foot/3m pole planted in the ground as a warning that one is entering "wild man" territory.

Left: Another Kwakiutl "buck'was" or "Wild Man of the Woods" mask. Right: Haida "gagit" or "Land Otter Man" mask (man-like creature that may be associated with the sasquatch). The "spines" in the lips are representations of sea urchin and fish dorsal spines which the gagit suffered in eating such food. Both masks are copies (replicas). It is believed the originals are very old.

Carved images in a dying tree, Salt Fork State Park, southeastern Ohio. Robert Morgan took this photograph in 1999. Upon encountering an elderly gentleman who was attending a family reunion on the site of his original homestead (his family lost their land to the Salt Fork dam) Morgan asked about the carvings. The gentleman told Morgan that the carvings represented a "Wooly-Boger," (another name for sasquatch). He further stated that as a child he had seen these creatures on many occasions.

FIRST NATIONS BIGFOOT PICTOGRAPHS

Pictographs, because they are surface paintings utilizing natural pigments such as chalk for white, charcoal for black and ochre for red, are more delicate and temporal than petroglyphs. Nevertheless, many sites protected from weathering, such as caves, have long lasting graphic representations of primitive life, including depictions of sasquatch-like creatures.

ASIDE: In September 2003 I attended a bigfoot symposium in Willow Creek, California. One of the symposium presenters, Kathy Moskowitz, provided an outstanding talk on First Nations pictographs that specifically depict bigfoot or sasquatch creatures. I had not seen evidence of this nature before and believe it to be highly important in the field of bigfoot/sasquatch studies. Most certainly, Kathy's findings are another significant indicator of the creature's reality. The following is a special presentation Kathy has kindly provided specifically for this work. CLM

"Most certainly, Kathy's findings are another significant indicator of the creature's reality."

Mayak datat: An Archaeological Viewpoint of the Hairy Man Pictographs located at Painted Rock, California

Kathy Moskowitz, U.S. Forest Service

Painted Rock is located on the Tule River Indian Reservation, above Porterville, in the Sierra Nevada foothills of central California. This site, also known as CATUL-19, is a rockshelter associated with a Yokuts Native American village. The site, located immediately adjacent to the Tule River, includes bedrock mortars, pitted boulders, midden and pictographs. The pictographs are located within the rockshelter, and are painted on the ceiling and walls of the shelter. The pictographs include paintings of a male bigfoot, a female bigfoot, and a child bigfoot (known as the family), coyote, beaver, bear, frog, caterpillar, centipede, humans, eagle, condor, lizard and various lines, circles, and other geometric designs. The paintings are in red, black, white, and yellow. All of these paintings are associated with the Yokuts creation story in which Hairy Man determined that people would walk on two legs.

The Yokuts Tribe occupied the San Joaquin Valley and foothills of California. The band of the Yokuts that lived at Painted Rock were called the Oching'-i-ta, meaning the "People of Painted Rock." A village at Painted Rock was called Uchiyingetau,

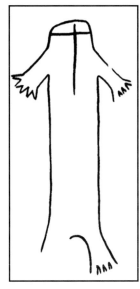

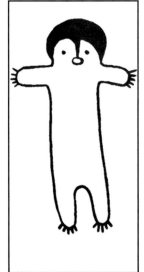

which means "markings." This implies the paintings were already there when the village was established. Painted Rock itself was called Hocheu. These names were recorded in 1877. Based on archaeological evidence, about 300 individuals occupied the village at Painted Rock year-round, and all aspects of village life, such as ceremonies, were conducted there.

The most dominant pictograph at the archaeological site is that of the Hairy Man, also known as mayak datat (mi!yak datr!atr!) or sunsunut (shoonshoonootr!). Hairy Man measures 8.45-feet/2.54m high, and is red, black, and white. The pictograph represents a two-legged creature, with its arms spread out 6-feet/1.8m across. It has what appears to be long hair and large haunting eyes. The Yokuts identify the lines coming from the eyes as tears (because Hairy Man is sad according to their creation story). The pictograph is in very poor condition due to weathering and vandalism. A Hairy Man petroglyph (something pecked into the stone, rather than painted) is present at the site as well. Petroglyphs are very rare in the Sierras (see photograph in the previous section).

Probably the most unusual feature of this site is the presence of an entire bigfoot family. Besides the male bigfoot (Hairy Man), there are also a female and child bigfoot. The mother is 5.85-feet/1.78m high by 3.9-feet /1.19m wide, and is solely red. The painting represents a two-legged female creature with her arms open. She has five fingers but other details are lacking. Immediately adjacent to her, and directly under her right hand, is her child. The child measures 3.9-feet/1.2m high by 3.25-feet /.99m wide and is also solely red. The painting represents a two-legged creature with small arms and five fingers. As far as can be determined, there are no other known bigfoot pictographs or petroglyphs in California.

The Yokuts have many stories involving the Hairy Man. Research by ethnographers has noted that Yokuts routinely incorporated direct observations of animal behavior into their traditional stories. The more they observed, the more elaborate their stories and details. Since Hairy Man, or bigfoot, is very prominent in their stories, much can be inferred about the creature's possible behavior.

As noted before, the Yokuts creation story attributes the ability of humans to walk on two legs to the Hairy Man. Although Hairy Man was pleased that he had helped create humans, "people" were afraid of his size and appearance and ran away from him. A second story, called *When People Took Over*, records that because people had spread over all the earth, animals had to find other places to live. Hairy Man says, "I will go live among the big trees (Giant Sequoias) and hunt only at night when people are asleep." A story called *Food Stealing* noted that Hairy Man was drawn by the sound of women pounding acorns in bedrock mortars (which sounds very much like woodknocking). He would wait for the women to process the bitter acorn meal before stealing it. A story called *Bigfoot, the Hairy Man*, talks about him eating animals (or people, if necessary), hanging out at the river, and generally having a sinister nature. He is also known to whistle to lure people outside.

The stories point out several behaviors or characteristics of a bigfoot. He is nocturnal, hunts and eats animals, is an omnivore, prefers forest environments, whistles, and may knock on wood to emulate acorn pounding. The pictographs clearly describe the physical characteristics attributed to Hairy Man (8.5-feet/2.6m tall, long shaggy hair, walks on two feet, and has a large, powerful, human-like body type). Taken together, and with the knowledge that the Yokuts incorporated direct observations of real animals into their stories and paintings, it is reasonable to assume that details on how a bigfoot looked and behaved are only present in Yokuts culture because of direct observation of a flesh and blood creature.

About Kathy Moskowitz

Kathy Moskowitz is currently the Forest Archaeologist for the Stanislaus National Forest, headquartered in Sonora, California. She is the primary person responsible for all archaeological and paleontological resources in her forest, but also directs education and public participation programs.

Kathy received a Bachelor of Arts degree in Anthropology in 1990 and a Master of Arts in Behavioral Science (emphasis in Anthropology) in 1994. Her main research interest involves prehistoric human ecology.

Kathy became interested in bigfoot as a child, and her interest lead her into the field of anthropology. In 1991, as an archaeologist for the Sequoia National Forest, she began interviewing elders from the Tule River Indian Tribe about their traditional Hairy Man stories. Since that time, she has gathered dozens of similar stories from various tribes in California and the Great Basin. As presented here, she has also conducted research on the Painted Rock Bigfoot Pictographs, the only known prehistoric "paintings" of bigfoot.

Kathy continues to research the connection between the traditional stories of hundreds of Native American tribes and bigfoot. She also feels that by studying the local environmental adaptations of prehistoric people, we might gain a greater insight into how bigfoot has similarly adapted itself to its environment. Similarities may be found in habitation methods and locations, hunting and gathering techniques, resource availability, calorie input/output, seasonal movements, etc. Knowing this information may foster a collection of evidence not previously associated with bigfoot behavior. It may also allow us to develop techniques to better obtain direct observation, and therefore documentation and protection, of this currently unrecognized primate.

Other photographs at the Painted Rock site showing a caterpillar, a coyote eating the moon and three people. The fact that other creatures and what might be considered "normal" people are depicted lends credibility to the conclusion that the "Hairy Man" was a totally different creature.

SASQUATCH IN NORTHWEST COAST NATIVE MYTHOLOGY

In native mythology sasquatch, like other animate and inanimate objects possessed the ability to "transform" from the real to the spiritual world and return.

As with many large and powerful creatures, the sasquatch appears in numerous legends told around the night time fires and is often involved in stories of kidnapping and great bravery. Elaborately carved masks used in ceremonies could be manipulated by pulleys to "transform" or reveal the different characters being represented in the stories.

This Kwakiutl D'sonoqua mask seen here is believed to represent the sasquatch. The mask with its associated robe was danced by Tsungani at the Chief Don Assu Potlatch in Campbell River, B.C. in 2002 as seen in the photograph on the opposite page.

Contemporary Kwakiutl D'sonoqua design.

Two transformation figures "revealed" by the D'sonoqua dancer seen on the opposite page.

Contemporary sasquatch doll by Lalooska.

D'sonoqua pole in Seattle, Washington.

Early Written Records

Early written references and recorded sightings that could refer to sasquatch creatures in North America go back about 200 years. Journals of early explorers, travelers, old newspapers and magazines carry reports of strange creatures that generally fit the description of a sasquatch. As can be expected, there are not many written reports in the early years. There were fewer people then and access to the media, as it were, was highly limited. Further, news of this nature was not "big news" so we can reason that many reports were probably ignored.

EARLY EXPLORERS AND TRAVELERS

The vast unexplored regions of North America were a formidable challenge to early explorers and travelers. Undoubtedly many journals, diaries and other writings are lost in history. Nevertheless, among those that have survived, the following recorded accounts of possible sasquatch related incidents are among the most noteworthy.

The explorer and geographer David Thompson (1770-1857), found unusual 14-inch/35.6cm, four-toed, clawed footprints near the present site of Jasper, Alberta in the winter of 1811. He does not state that the tracks appeared to have been made by a creature with four legs or two legs.

However, as the First Nations people in his party would not acccept that the tracks were made by a bear, then we have a little mystery. Some researchers believe what he saw were sasquatch tracks, but sasquatch prints generally show five toes and no claws. Nevertheless, other alleged sasquatch prints showing only four toes have been found and some prints indicated claws or nails. The Canadian postage stamp shown was issued in 1957. There is no known painting of Thompson.

© Canada Post Corp., 1957, reproduced with permission

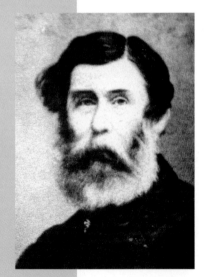

Paul Kane

The noted explorer and artist Paul Kane also references unusual creatures in his book, *The Wanderings of an Artist*. In his entry for the date March 26, 1847, Mt. St. Helens area, Washington, he states, *This mountain has never been visited by either Whites or Indians; the latter assert that it is inhabited by a race of beings of a different species, who are cannibals, and whom they hold in great dread; they also say that there is a lake at its base with a very extraordinary kind of fish in it, with a head more resembling that of a bear than any other animal. These superstitions are taken from a statement of a man who, they say, went to the mountain with another, and escaped the fate of his companion, who was eaten by the "Skoocooms," or evil genii. I offered a considerable bribe to any Indian who would accompany me in its exploration, but could not find one hardy enough to venture."*

Remarkably, the first major published report of a possible sasquatch encounter is in a book entitled Wilderness Hunter *(1892) by Theodore Roosevelt, who later became President of the United States. In his book, Roosevelt provides a very detailed account of a story he was told by a hunter named Bauman. As the story goes, Bauman's trapping companion was viciously killed by a "beast creature" that walked on two legs. Roosevelt heard the story while he was in the Bitterroot Mountains which are on the Idaho-Montana border. By this time, Bauman was an old man and the incident he related probably took place in the late 1850s.*

EARLY NEWSPAPER REPORTS

Early newspaper reports generally refer to unusual man-like creatures as "wild men." Indeed, it can be reasoned that many such creatures were exactly that. From all of the early articles I have read, the following are among the most credible with regard to possible sasquatch creatures.

This article appeared in the *Memphis Enquirer* on May 9, 1851.

WILD MAN OF THE WOODS: *During March last, Mr. Hamilton of Greene County, Arkansas, while out hunting with an acquaintance, observed a drove of cattle in a state of apparent alarm, evidently pursued by some dreaded enemy. Halting for the purpose, they soon discovered as the animals fled by them, that they were followed by an animal bearing the unmistakable likeness of humanity. He was of gigantic stature, the body being covered with hair and the head with long locks that fairly enveloped his neck and shoulders. The "wildman," for so we must call him, after looking at them deliberately for a short time, turned and ran away with great speed, leaping from twelve to fourteen feet at a time. His footprints measured thirteen inches each. This singular creature has long been known traditionally in St. Francis, Greene and Poinsett Counties. Arkansas sportsmen and hunters having described him so long as seventeen years since. A planter indeed saw him very recently, but withheld his information lest he should not be credited, until the account of Mr. Hamilton and his friend placed the existence of the animal beyond cavil." A party was to leave Memphis in pursuit of the creature.*

"He was of gigantic stature, the body being covered with hair and the head with long locks that fairly enveloped his neck and shoulders."

In 1884, an intriguing article appeared in *The Colonist*, a Victoria, British Columbia newspaper. The article states that a creature, "something of the gorilla type" had been captured near Yale, British Columbia. The photograph seen here shows Yale in about 1880. The creature is described as standing about 4-feet, 7-inches/1.4m in height and weighing 127-pounds/57.5kg. From these measurements, we might conclude that it was a young sasquatch. The following is an exact reprint of the article.

What Is It?
A Strange Creature Captured Above Yale
A British Columbia Gorilla
Correspondence to The Chronicle

"He has long, black, strong hair and resembles a human being with one exception, his entire body, excepting his hands, (or paws) and feet are covered with glossy hair about one inch long."

Yale, B.C., July 3, 1884.[1] In the immediate vicinity of No. 4 tunnel, situated some twenty miles above this village, are bluffs of rocks which have hitherto been unsurmountable, but on Monday morning last were successfully scaled by Mr. Onderdonk's employees on the regular train from Lytton. Assisted by Mr. Casterton, the British Columbia Express Company's messenger, and a number of gentlemen from Lytton and points east of that place who, after considerable trouble and perilous climbing, succeeded in capturing a creature which may truly be called half man and half beast. "Jacko," as the creature has been called by his captors, is something of the gorilla type standing about four feet seven inches in height and weighing 127 pounds. He has long, black, strong hair and resembles a human being with one exception, his entire body, excepting his hands, (or paws) and feet are covered with glossy hair about one inch long. His fore arm is much longer than a man's fore arm, and he possesses extraordinary strength, as he will take hold of a stick and break it by wrenching or twisting it, which no man living could break in the same way. Since his capture he is very reticent, only occasionlly uttering

---

[1] *The actual newspaper article shows the date as 1882. This is an obvious error that I have corrected. CLM*

a noise which is half bark and half growl. He is, however, becoming daily more attached to his keeper, Mr. George Tilbury, of this place, who proposes shortly starting for London, England, to exhibit him. His favorite food so far is berries, and he drinks fresh milk with evident relish. By advice of Dr. Hannington raw meats have been withheld from Jacko, as the doctor thinks it would have a tendency to make him savage. The mode of capture was as follows: Ned Austin, the engineer, on coming in sight of the bluff at the eastern end of the No. 4 tunnel saw what he supposed to be a man lying asleep in close proximity of the track, and as quickly as thought blew the signal to apply the breaks. The brakes were instantly applied, and in a few seconds the train was brought to a standstill. At this moment the supposed man sprang up, and uttering a sharp quick bark began to climb the steep bluff. Conductor R. J. Craig and Express Messenger Custerton, followed by the baggageman and brakesmen, jumped from the train and knowing they were some twenty minutes ahead of time gave immediate chase. After five minutes of perilous climbing the then supposed demented Indian was corralled on a projecting shelf of rock where he could neither ascend or descend. The query now was how to capture him alive, which was quickly decided by Mr. Craig, who crawled on his hands and knees until he was about forty feet above the creature. Taking a small piece of loose rock he let it fall and it had the desired effect on rendering poor Jacko incapable of resistance for a time at least. The bell rope was then brought up and Jacko was now lowered to terra firma. After firmly binding him and placing him in the baggage car "off brakes" was sounded and the train started for Yale. At the station a large crowd who had heard of the capture by telephone from Spuzzum Flat were assembled, each one anxious to have the first look at the monstrosity, but they were disapointed, as Jacko had been taken off at the machine shops and placed in charge of his present keeper.

The question naturally arises, how came the creature where it was first seen by Mr. Austin? From bruses about its head and body, and apparent soreness since its capture, it is supposed that Jacko ventured too near to the edge of the bluff, slipped, fell and lay where found until the sound of the rushing train aroused him. Mr. Thos. White and Mr. Gouin, C.E., as well as Mr. Major, who kept a small store about half a mile west of the tunnel during the past two years, have mentioned having seen a curious creature at different points between Camps 13 and 17, but no attention was paid to their remarks as people came to the conclusion that they had either seen a bear or stray Indian dog. Who can unravel the mystery that now surrounds Jacko? Does he belong to a species hitherto unknown in this part of the continent, or is he really what the train man first thought he was, a crazy Indian?

Older residents of Yale interviewed many years later recalled stories of the incident and one resident stated that his grandfather actually saw the creature. Nevertheless, the story could have been fabricated — hoaxes were commonly practiced at that time.

WHAT HAPPENED TO JACKO?

The last we know of Jacko is that he was shipped in a cage to England to be used in a sideshow, but apparently never arrived at that destination.

Across the continent, also in the year 1884, the Barnum & Bailey Circus presented in New York City Jo-Jo the Dog-Faced Boy, a sixteen-year-old youth covered in long hair. Jo-Jo, whose actual name was Fedor Jeftichew (b. 1868), was alleged to have been found in Russia along with his father who was also covered in hair.

Jo-Jo has been coincidentally connected with Jacko. It has been reasoned that Jacko may have been purchased in the U.S. by circus man P. T. Barnum and billed for a side-show but died before he could be exhibited. Barnum thereupon quickly found a replacement — Jo-Jo. Circus advertising material created in 1884 showing a hairy creature does not appear to show Jo-Jo. The material was replaced with an actual photograph of Jo-Jo taken in 1885.

The Sasquatch "Classics"

With improved communications and more people (prospectors, hunters, campers, road construction workers, etc.) pushing into North America's wilderness, reports of sasquatch sightings and discoveries of unusual footprints dramatically increased. We can reason, however, that the number of incidents reported is only a fraction of the actual number. In all probability most incidents, for a variety of reasons, are not publicized. Nevertheless, reports of sasquatch sightings and footprint findings during the twentieth century number in the thousands. Six reports prior to the 1960s emerge as the "classics." All continue to be examined and researched with on-going debate.

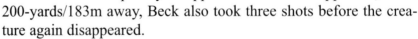
FRED BECK & THE APEMEN OF MT. ST. HELENS

Fred Beck

In the summer of 1924, Fred Beck, seen here with his trusty rifle, and four other prospectors state they were attacked by a number of sasquatch. The men had been prospecting in the Mt. St. Helens and Lewis River area, southern Washington State, for about six years. They had staked a gold claim, which they named the Vander White, about 2-miles/3.2km east of Mt. St Helens. Here, they built a cabin near a deep canyon. Occasionally, they saw large footprints, which as far as they knew did not match those of any known animals. The largest print they observed measured 19-inches/48.3cm. One evening they heard peculiar whistling and thumping sounds that continued for about one week. Later, while Beck and one of the other men were getting water at a nearby spring, they observed a strange creature about 100-yards/91.4m away. The other man took three rifle shots at the creature which quickly disappeared. When it reappeared, about 200-yards/183m away, Beck also took three shots before the creature again disappeared.

After the other men were informed of this incident, all agreed to go home the following morning. That night, however, several of the creatures attacked the cabin. They started by pelting the cabin with rocks. As there were no windows in the cabin, the men could not see the assailants. The men's only view outside was through a chinking space. With the limited field of view and the darkness, nothing was actually seen. Later the creatures climbed on the cabin roof, and tried to break down the door. The men fired their rifles through the roof and through the door. One creature even reached into the cabin through the chinking space and grabbed hold of an ax. Marion Smith, Beck's father-in-law, turned the ax head so that it caught on the logs. He then shot his rifle along the ax handle and the creature let go of the ax. The attack ended just before daylight. When it was light enough, the men ventured outside. A short time later, Beck saw one of the creatures about 80-yards/73.9m away near the edge of the canyon. He took three shots at it and saw it topple into the gorge which was about 400-feet/122m deep. The men hastily left the area without packing their supplies and equipment. They took only what they could carry in their packsacks.

In 1967, Fred Beck wrote a small book on his experience. The drawing is by Everett Davenport.

The top end of Ape Canyon, so named for the unusual encounter between a number of ape-like creatures and Fred Beck with his party of miners in 1924.

ALBERT OSTMAN'S INCREDIBLE JOURNEY

NOTE: This "classic" is so bizarre it defies the imagination. Nevertheless, it has been tightly woven into sasquatch lore so needs to be presented.

During the summer of 1924, Albert Ostman, a construction worker, went to look for gold at the head of Toba Inlet, British Columbia. After a two-day trek, he set up his permanent campsite. When he awoke the next morning, he found that his things had been disturbed, although nothing was missing. He was a heavy sleeper, so was not surprised that he slept through the intrusion. The next morning he awoke to the same thing, but this time his packsack had been emptied out and some food was missing.

After a third "visit," Ostman determined to stay awake all night to catch the intruder. He climbed into his sleeping bag fully clothed, save his boots, with his rifle by his side in the bag. He placed his boots at the bottom of the bag. He fell asleep but was then awakened by something picking him up and carrying him, sleeping bag and all. He was bundled up in such a way that he could not move. Whatever was carrying him also had his packsack, as Ostman could feel food cans touching his back. Having heard of "mountain giants," Ostman reasoned that it was one of these creatures carrying him. After a journey of some three hours, the creature unloaded his cargo onto the ground. Upon climbing out of the bag and getting himself together, Ostman discovered he was in the company of four sasquatch, two adults and two children.

Ostman spent six days with his captors and was able to observe firsthand (and later recount in considerable depth) how the creatures looked and lived. He then made his escape by tricking the adult male into eating a box of snuff (chewing tobacco).

While Ostman's story is amusing and interesting, there are obvious details that detract from the story's credibility. For that reason, I prefer to refrain from providing a full account of his observations. Certainly the story is possible. We have one of those situations were the information appears to be too good to have been fabricated but such does not eliminate this possibility.

The first page of Albert Ostman's scribbler.

"Upon climbing out of the bag and getting himself together, Ostman discovered he was in the company of four sasquatch, two adults and two children."

Albert Ostman, right, is seen here being interviewed by René Dahinden. Ostman is holding the scribbler in which he wrote his unusual story.

JOHN W. BURNS & THE CHEHALIS FIRST NATIONS PEOPLE

Certainly, the Chehalis First Nations people in British Columbia measure significantly in sasquatch lore. Legend has it that these people once fought a great battle with the giant creatures. In 1980, a sasquatch figure, as seen here, was adopted as the symbol of the Chehalis First Nations Band in B.C.

The following article that appeared in *Liberty* magazine in December 1954 gives us a good appreciation of the Chehalis/sasquatch connection. The person who wrote the article, John W. Burns, developed the word "sasquatch" which is now the common name for the creature in Canada.

MY SEARCH FOR B.C.'S GIANT INDIANS
by JOHN W. BURNS *as told to Charles V. Tench*

Do the hairy, 8-feet tall Sasquatch still live? I have spent over 16 years, as a teacher at Chehalis Indian Reserve, seeking them.

I have spent more than 16[1] years trying to track down in the unexplored wilds of British Columbia, Canada's most elusive tribe of Indians. They are the mysterious Sasquatch - wild giants eight feet tall, covered from head to toe with black, woolly hair.

My search for these primitive creatures began in 1925 when, after serving on the Vancouver Sun, I was appointed teacher for the Chehalis Indian Reserve. Here, buried in the bush by the banks of the Harrison River, B.C., some 60 miles from Vancouver, my wife and I have been friends for 16 years with the Chehalis Indians.

Because they knew I wouldn't taunt them, my Chehalis neighbors revealed to me the secrets of the Sasquatch - details never confided to any white man before. The older Indians called the tribe "Saskehavis," literally "wild men." I named them "Sasquatch," which can be translated freely into English as "hairy giants." I've never personally encountered a Sasquatch myself. Yet I've compiled an imposing dossier of first-hand accounts from Indians who have met the wild giants face to face and know survivors of the tribe still live today. I was always aware when the Sasquatch were in the vicinity of our Indian village, for then the children were kept indoors and not allowed to venture to my school. The Chehalis Indians are intelligent, but unimaginative, folk. Inventing so many factually detailed stories concerning their adventures with the giants would be quite beyond their powers.

Certainly, they are highly sensitive when white strangers ridicule their well-authorized stories. Once, on May 23 and 24, 1938, an "Indian Sasquatch Days" festival was held at Harrison Hot Springs,

> "Because they knew I wouldn't taunt them, my Chehalis neighbors revealed to me the secrets of the Sasquatch - details never confided to any white man before."

> "I named them *Sasquatch*, which can be translated freely into English as *hairy giants*."

---

[1] It appears Burns wrote his article in 1941. We know he was still at Chehalis until 1945.

B.C. After getting special permission from the Department of Indian Affairs, Ottawa, I took several hundred of my Indians.

Unhappily, a prominent member of the B.C. Government made a hash of the ceremonies. In his welcoming speech over the microphone, the official blundered: "Of course, the Sasquatch are merely Indian legendary monsters. No white man has ever seen one. They do not exist today. In fact …"

He was drowned out by a rustling of buckskin garments and tinkling of ornamental bells as, in response to an indignant sign from old Chief Flying Eagle, over 2,000 Indians rose to their feet in angry protest. The Chief stalked to the open space where the government officials stood, and, turning his back on them, thundered into the mike in excellent English:

"The speaker is wrong! To all who now hear, I, Chief Flying Eagle say: Some white men have seen Sasquatch. Many Indians have seen Sasquatch and spoke to them. Sasquatch still live all around here. Indians do not lie!"

Ever since my interest in the Sasquatch was stimulated by the celebrated anthropologist, Prof. Hill Tout, I've come across fascinating proof. Oldest written record I discovered was that of the late Alexander Caulfield Anderson, after whom the West Vancouver suburb, Caulfield, is named. When he was a Hudson's Bay Co. inspector in 1846, establishing a post near Harrison Lake, Anderson frequently mentioned in his official reports "the wild giants of the mountains." Once, he wrote, he and his party were met by a bombardment of rocks hurled by a number of Sasquatch.

What do the modern Sasquatch look like? I was given a vivid description by William Point and Adaline August, Indian graduates of a Vancouver high school. They encountered a wild giant last September, four miles from the picnic that Indian hoppickers hold annually near Agassiz, B.C.

"We were walking on the railroad track toward the house of Adaline's parents," Point told me, "when Adaline noticed a person coming toward us. We halted in alarm. The man wore no clothing at all, and was covered with hair, like an animal."

"He was twice as big as the average man. His arms were so long his hands almost touched the ground. His eyes were large and fierce as a cougar's. The lower part of his noise was wide and spread over the greater part of his face, which gave him a repulsive appearance."

"Then my nerve failed me. I turned and ran."

The Indians tell me that each summer the Sasquatch have a gathering of the survivors of their race near the rocky, shelving top of Morris Mountain. Just before the reunion, the giants send out scouts. It's these scattered scouts that Chehalis Indians have met.

Naturally, reports of the giants have drawn the interest of anthropologists. Two years ago, an American expedition, equipped with movie cameras, asked me to enlist the aid of Indian guides. Though offered $10 a day, not one of my Indians would volunteer.

"It would be in vain," the Chehalis said. "The Sasquatch, seeing the expedition approach, would immediately go into hiding."

The American party set out without native guides. In two weeks, they returned, weary and fly-bitten.

"For an ordinary white man," they told me, "the way to the top of Morris Mountain is utterly impossible."

Yet I have accepted all the Sasquatch encounters recounted to me in good faith. One Indian known for his truthfulness, Peter Williams, told me he was chased and almost had his frame shack pushed over by a wild giant in the Saskahaua, or "Place of the Wild Men," district of B.C. Next morning, Peter measured the giant's tracks in the mud. The footprints were 22 inches long - compared with the average man's 10 to 12-inch tracks.

Another Indian in a canoe, Chehalis Phillip, had a rock hurled at him by a hairy giant. One of my Indians, Charley Victor, wounded a 12-year old naked giant living in a tree trunk, and was scolded by a seven-foot Sasquatch woman in the Douglas dialect: "You hurt my friend!"

But perhaps the strangest experience happened to a Chehalis woman, Serephine Long. She told me she was abducted by a Sasquatch and lived in the haunts of the wild people for about a year. Just before she was about to marry a young brave named Qualac (Thunder), while she was gathering cedar roots, a hairy young giant leaped on her from a bush. He smeared tree gum over her eyes so that she couldn't see, hoisted her to his shoulder, and raced off with the struggling woman to a cave on Mount Morris.

There she was kept prisoner, living with the Sasquatch and his elderly parents. "They fed me well," she said.

After almost 12 months, she grew sick and pleaded, "I wish to see my own people before I die." Her young Sasquatch reluctantly put tree gum on her eyelids once more and carried her back.

"I was too weak to talk to my people when I stumbled into the house," she recalled to me. "I crawled into bed and that night gave birth to a child. The little one lived but for a few hours, for which I was glad. I hope that never again shall I see a Sasquatch."

Many of my other Indians are sincerely convinced the Sasquatch live in the unexplored interior of B.C. And with the Indians, whom I know and trust, I also believe.

Your author and René at the Chehalis reserve. It is beautiful wilderness in which one has no problem understanding how a creature such as the sasquatch could survive and stay hidden from human eyes.

The Chehalis River on a sunny day in 2003.

The Chehalis administrative and community building.

Brad Tombe is seen here with two casts he made of prints he found near the Chehalis River.

Chehalis Revisited

While I was visiting René Dahinden in the summer of 1995, a young fellow by the name of Brad Tombe came by and showed us a cast of a footprint (one of several) he found by the Chehalis River. Brad gave me a written report (seen below), however, René was not too impressed and wrote the cast off as a bear print. A short time later, he received a telephone call from the Chehalis Chief, Alexander Paul, informing of a sasquatch sighting by one of his people. René, my son Dan, and I went up to investigate the next day. Chief Paul (left) is seen here with Dan.

We interviewed the witness who stated that he saw what he first thought was a bear walking in the river. However, as he watched, it stood up and walked out of the river on two legs. Apparently the creature had been bending over, probably in the process of obtaining food of some sort. We went out and searched the whole area but could not find anything. I recently (April, 2002) drove up to the reservation. Logging operations were in full swing with a muddy logging road cutting deep into the forest. The Chief was not available during my visit so I telephoned him the next day. He informed me that in the last three years there have been three more sasquatch sightings in the Chehalis area. It would not surprise me if the Chehalis River were to become the Bluff Creek of British Columbia.

```
Location: Chehalis River
Weather:  Rainy/Overcast
Date:     August, 6, 1995
Time:     3:30pm
Notes:    I walked down river and fished the runs as I went. The river
bank was quite rocky and in places sand occasionally appeared in small
stretches. When I walked past one of these patches it appeared to have
some sort of tracks through it. As I began to look at them the tracks
appeared to be footprints. At the time I was not sure what they were
but it looked like a large human foot. One could clearly see a heel
and large mound of sand that had been pushed up. The front of the foot
could also be seen and it appeared that a large toe was present. I
decided that I would practice my plaster skills and poured out a few
casts. When I proceeded to do so it was then that I could then see the
shape of the footprint. Another angler stopped and helped me measure
what appeared to be the stride of the person and it was 50"{4 Feet}.
There were four tracks and all were 12" in length and I photographed
them all with the measuring tape beside them for comparison.

               Brad Tombe
```

Brad Tombe's report which he gave me when he visited René Dahinden.

THE RUBY CREEK INCIDENT

George Chapman (First Nations) lived with his wife, Jeannie, and their three children (8, 7, and 5 years old) in a small isolated house on the banks of the Fraser River near Ruby Creek, British Columbia. George was employed as a railroad maintenance worker at Ruby Creek. In September 1941, he was surprised to see his wife and children running down the tracks towards him. Jeannie excitedly told her husband that a sasquatch was after her. As it happened, one of the children was playing in the front yard of the house. The child came running into the house shouting that a "big cow" was coming out of the woods. Jeannie looked out and saw an ape-like creature, 7.5-feet/2.3m tall, covered in dark hair, approaching the house. Terrified at the sight of the creature, she grabbed her children and fled.

George and other men went to the house and found 16-inch /40.6cm footprints that led to a shed where a barrel of fish had been dumped out. The prints then led across a field and into the mountains. Footprints on each side of a wire fence, 4 to 5-feet (1.2-1.5m) high, gave another clue as to the size of the creature – it apparently just took the fence in stride.

The incident was thoroughly investigated by researchers and a cast was made of one of the creature's foot prints (cast length was 17-inches/43.2cm). The Chapmans returned to their home but were continually bothered by unusual howling noises and their agitated dogs (which appeared to sense an unusual "presence"). The family left the house within one week and never returned.

Tracing of the 17-inch/43.2cm footprint cast made from a print left by the creature on the Chapman's property. The tracing was made by Deputy Sheriff Joe Dunn of Bellingham, Washington. Unfortunately the actual cast eventually shattered and was discarded.

This is the view from the front of the house in 1995.

René Dahinden is seen here on the fence over which the creature merely stepped.

For many years the Chapman house remained vacant and eventually crumbled. These photographs show the house probably fifteen years or so after it was left to the elements.

THE WILLIAM ROE EXPERIENCE

"Roe leveled his rifle at the creature to kill it. However, he changed his mind because he felt it was human."

In October 1955, William Roe, a highway worker and experienced hunter and trapper, decided to hike up to a deserted mine on Mica Mountain, which is near Tete Jaune Cache, British Columbia.

Roe was working on the highway near the town and decided on a hike for something to do. Just as he came within sight of the mine, he spotted what he thought was a grizzly bear half hidden in the bush about 75-yards/68.6m from where he was standing. He had his rifle with him but did not wish to shoot the animal as he had no way of getting it out. He therefore calmly sat down on a rock behind a bush and observed the scene. A few moments later the animal rose up and stepped out into the open. He now saw that it definitely was not a bear, but what appeared to be a man-like creature, about 6-feet/1.8m tall, covered in dark brown silver-tipped hair.

The creature, unaware of Roe's presence, walked directly towards him. Row then observed by its breasts that the creature was female. It proceeded to the edge of the bush where Roe was hiding, within twenty feet of his position. Here it crouched down and began eating leaves from the bush, remaining for a considerable time before realizing it was being watched. During all of this time Roe was able to observe many important details as to how the creature walked, its physical makeup and eating habit of drawing branches through its teeth. When the creature noticed Roe, a look of amazement crossed its face which Roe found comical and chuckled to himself. Remaining crouched, the creature then backed away three or four steps. It thereupon straightened up and rapidly walked away in the same direction from which it had arrived, glancing back once at Roe over its shoulder.

Realizing he had stumbled on something of great scientific interest, Roe leveled his rifle at the creature to kill it. However, he changed his mind because he felt it was human. In the distance, the creature threw its head back on two occasions and emitted a peculiar noise that Roe described as "half laugh and half language."

Roe's examination of feces in the area, which he believed was from the creature, convinced him that it was strictly a vegetarian.

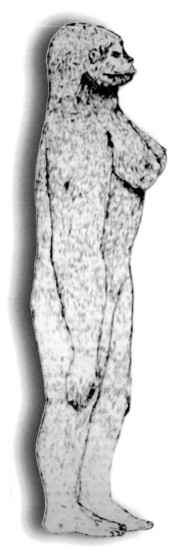

Drawing of the creature seen by Roe made by his daughter under his direction.

JERRY CREW & THE BIRTH OF THE NAME "BIGFOOT"

A road was constructed into the Bluff Creek, California region in 1957, opening the area, which up to that time had been remote wilderness. Over the years, people in the region had noticed large human-like footprints and while at least one report had been provided to a local newspaper, no attention was paid to the matter. On August 27, 1958, Gerald (Jerry) Crew a road construction worker, saw such prints circling his parked bulldozer. Crew had heard of similar findings by a road gang about one year earlier at a location eight miles north. He showed the prints to his fellow workers, some of whom said they had also seen prints in the area. Whatever was making the prints was appropriately being referred to by the men as "Big Foot." Crew saw additional prints about one month later and more on October 2, 1958. This time, he made a plaster cast of one of the prints. He took the cast to the *Humboldt Times* newspaper (Eureka, California) and related the story of his find. An Associated Press release on the story used the term "Bigfoot" which resulted in this name becoming the recognized name for the creature in the United States. The first photograph seen here showing Jerry Crew holding his cast was used for the article in the *Vancouver Province* newspaper (Vancouver, British Columbia, Canada) published on Monday, October 6, 1958. The second photograph was used by the *Humboldt Times* newspaper in its article (written by Andrew Genzoli) published on October 14, 1958.

Together with the footprints found at that time, there were also alleged sightings and other unusual incidents in the area. In later years, investigations revealed tracks of six different sizes, indicating that a number of bigfoot frequented the area. Footprint sizes ranged from 12.25-inches to 17-inches (43.2cm to 31cm) long. These facts made the Bluff Creek area a prime location for a possible bigfoot sighting.

Jerry Crew (seen here in both photographs) later moved to San Francisco where he worked with an airline company. He eventually settled in northern Oregon where he died in the fall of 1993.

Organized Expeditions to Find the Sasquatch

The only major fully-organized and funded attempt to find a sasquatch was the Pacific Northwest Expedition (PNE) which commenced operations in 1959 and continued for almost three years. It had men in the field steadily. Several sets of footprints were found along with possible hair and droppings. The organization was funded and headed by Tom Slick, a Texas oil millionaire, with a burning desire to find both the sasquatch and yeti. Everything the researchers found including all photographs was sent to Slick's research facility at San Antonio, Texas. Unfortunately, Slick was killed in 1962 when his private plane shattered in the air over Montana. Slick's associates apparently did not share Tom's interests and disposed of all material. To our knowledge, hair and droppings were not positively identified, although we must remember that technology in the 1950s was far removed from current technology. Tom Slick also initially financed the British Columbia Expedition, headed by Bob Titmus, which started in 1961. After Slick died, Titmus carried on himself using his own finances until he could no longer afford to do so. The following is Johns Green's recollection of the events as related to me.

Willow Creek, California, 1960. It was here that the PNE was formed in the fall of the previous year. Little did the organizers know that 8 years later Roger Patterson and Bob Gimlin would film a sasquatch fewer than 50 -miles/80 km northwest of the little town.

While the PNE was the first organized expedition, there were people investigating sasquatch reports in B.C. in the first half of the twentieth century, including journalist/historian Bruce McKelvie; onetime North Vancouver mayor Charles Cates, and pre-eminently, John W. Burns, teacher and Indian agent on the Chehalis reserve beside the Harrison River. It was Burns whose writings introduced the name and the subject to British Columbians in the pages of the *Vancouver Daily Province* and to all Canadians in *Maclean's* magazine. There was also at least one serious investigator south of the U.S. border, Deputy Sheriff Joe Dunn, from Bellingham. By the time the tracks of "bigfoot" started turning up in northern California in 1958, however, the only people active in the field were René Dahinden and John Green, and the only one in California who took up the investigation in a major way after 1958 was Bob Titmus, who then had a taxidermy business near Redding, California.

In the fall of 1959 Titmus, Dahinden and Green were seeking funding to be able to spend more time in the quest, and through British zoologist Ivan Sanderson they made contact with Tom Slick. Meeting at Willow Creek, in California, the four men worked out a deal for Slick to put up some money (initially $5,000) for what he insisted on calling an "expedition," of which he was to have the title of "leader," while Titmus was to be "deputy leader," in charge of field operations.

That was the start of the now almost legendary "Pacific Northwest Expedition." Slick wanted a bigfoot hunted down with hounds, and supplied rifles for that purpose, while Bob hired a series of local hound hunters. Since tracks had been turning up fairly frequently, it seemed reasonable to suppose that success was only a matter of time, and not much of that, but experience proved otherwise. Some tracks were found and photographed, but none as good as those of which Bob had already made casts, and some samples of hair and of feces were sent to experts who were unable to identify them. In the summer of 1960 Titmus, Green and Dahinden had left the group. The expedition continued and everything collected had to be sent to Slick's "Southwest Foundation" in San Antonio. If anything was accomplished, no evidence of it was kept after Slick died.

The following year Sanderson received word of frequent sightings of what were called "apes" by First Nations people from Klemtu, a village on Swindle Island on the central B.C. coast. He and Green tried to find a different financial backer, but eventually had to turn to Slick, who paid for Green, Titmus and famed Bella Coola grizzly hunter, Clayton Mack, to fly to Klemtu and investigate. Prospects looked good, and a three-way deal was negotiated with Slick, again the "leader," of what was called the "British Columbia Expedition." Titmus, who had now sold his taxidermy business, became full-time "deputy leader," hunting by boat among the inlets and islands. At first it was a three-man effort, but Green could spend only one month each year, and soon it was just Titmus, often with a local helper. Slick joined in when he could, sometimes bringing his two young sons, but when he was killed in the summer of 1962 his associates quickly withdrew support. Titmus went back to California to wind up the PNE for them, and then he carried on hunting full time on the B.C. coast until his own money ran out and part time after that. He did find tracks, and had one sighting at a great distance, but the casts he made were all lost when his boat burned and sank. Thus, from Tom Slick's two "expeditions," little but the legend remains.

TOM SLICK

Tom Slick was born in 1916. His father, Tom Slick Sr., died in 1930, leaving behind his wife, two sons and a daughter. He had made a fortune in the oil business and his estate at the time of his death was reported to be in excess of $75 million. Young Tom went on to become a highly successful businessman himself and set up a series of research centers for the "betterment of humankind." In 1946, he and his brother founded Slick Airways.

Tom was highly interested in the "unexplained" and between 1956 and 1959 he led or sponsored several major expeditions to search for the yeti. Developments in North America in 1958 related to the sasquatch caught his attention and he subsequently funded a search for the creature.

John Green was a member of the Pacific North Expedition and took this photograph. The expedition members seen here from left to right are Ed Patrick, Tom Slick, René Dahinden, Kirk Johnson, Bob Titmus and Gerri Walsh (Slick's secretary).

The Patterson/Gimlin Film

5

On October 20, 1967, the course of events in sasquatch research took a dramatic new direction. It was on this day that Roger Patterson and Robert Gimlin filmed what they allege was an actual sasquatch creature. The photograph seen here is frame 364 of the film. A full presentation on the clearest frames follows.

Both men were residents of Yakima County, Washington State. Patterson became interested in sasquatch after reading an article on the creatures by Ivan T. Sanderson[1] in the December 1959 issue of *True* magazine. Patterson started active sasquatch research sometime in 1964. Gimlin was interested in the subject and accompanied Patterson on excursions to look for evidence of the creature. Patterson wrote a book entitled *Do Abominable Snowmen of America Really Exist?* that was published in 1966, about one year before he and Gimlin captured the creature on film. Patterson had considerable artistic talent and illustrated his book with drawings of sasquatch mainly in situations based on reported sightings of the creature. The bigfoot sculpture seen here was created by Patterson in the early 1960s.

[1] Sanderson was known as "a prominent British animal collector and zoologist."

The legendary Patterson book is presently being reprinted with the recent analysis of the film, including release for the first time, of additional support for the film by Chris Murphy. This is being published by Hancock House in 2004.

40

ACCOUNT OF THE FILMING ADVENTURE

At some point early in the year 1967, Patterson decided to make a film documentary on sasquatch. He subsequently rented a movie camera for this purpose. He wished to find and film fresh footprints as evidence of the creature's existence.

During late August and early September 1967, Patterson and Gimlin were exploring the Mt. St. Helens area. While they were away, Mrs. Patricia Patterson received word on the finding of large human-like footprints in the Bluff Creek, California area. She gave her husband this information when he returned home. It might be noted that this same area was the scene of considerable sasquatch activity nine years earlier. It was here in 1958 that Jerry Crew found large human-like footprints as previously related.

Upon receiving the information about the footprints from his wife, Patterson contacted Gimlin and the two made plans to investigate the findings. They traveled to the Bluff Creek area in Gimlin's truck, taking with them three horses. They departed on or about October 1, 1967. They set up their camp at Louse Camp which is located where Notice Creek and Bluff Creek join. The men then went out on horseback each day and explored the area. Patterson was intrigued with the scenery and autumn colors. He used 76-feet/23.2m of his first film roll for general filming, including shots of himself, Gimlin and the horses. During the evening, the men drove the rough roads hoping for a possible night sighting.

On the morning of October 20, 1967, Gimlin arose early and rode out of the campsite area while Patterson slept-in. Gimlin arrived back at the camp at about 10:00 a.m. Patterson was not at the camp at this time. He returned after a little while and asked Gimlin what area he had covered on his early ride. Gimlin told him where he had been after which Patterson suggested they re-explore an area they had previously explored. Gimlin agreed and the men left at about twelve noon.

At about 1:30 p.m. that day, Friday, October 20, 1967, Patterson and Gimlin spotted a female sasquatch down on a Bluff Creek gravel sandbar, across the creek from their position. Patterson estimated the creature to be about 6-feet/1.8m tall, maybe taller, and weighing about 400 pounds/181.2kg. Patterson's horse reared in alarm at the sight of the creature, bringing both horse and rider to the ground, Patterson pinned below. Gimlin's horse and the

BLUFF CREEK PRINTS

Prints found on or near the Blue Creek Mountain road (under construction) which is in the Bluff Creek area, had already been investigated by John Green alone and then by both Green and René Dahinden (late August 1967). Green requested the B.C. Provincial Museum to send someone down to see the prints. Don Abbott was sent and he saw the prints first hand. The road construction foreman held up work until he thought the team had finished. A misunderstanding here resulted in the prints being all but destroyed.

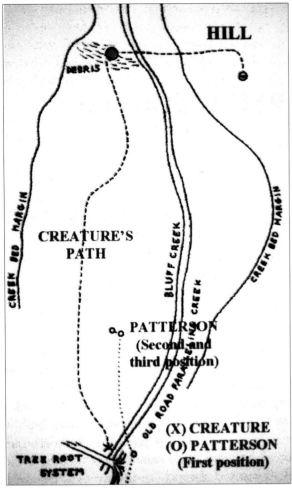

packhorse, being led by Gimlin, also reacted. The packhorse panicked and Gimlin released its lead in order to control the horse he was riding. Patterson, being an experienced horseman, quickly disengaged himself and grabbed his camera. He ran towards the creature, crossing the creek and stopping within about 80-feet /24.4m. While running, stationary, and later walking, Patterson took 24-feet/7.3m of color film footage which expired the film roll in the camera. During this time, the creature walked upstream into a sparsely wooded area where it disappeared from view. It does not appear that the creature was overly concerned with the men's presence (i.e., felt immediately threatened) until it turns and looks at Patterson - about one-third of the time into the filming.[2] After this point the creature hastened its pace somewhat as it continued its passage. In the meantime, Gimlin, on horseback, rode slowly towards the creature. Gimlin crossed the creek and dismounted. He then observed the whole scene, rifle in hand, in case his friend was attacked by the creature. The men had previously agreed that under no circumstances would they shoot a sasquatch unless to protect themselves or each other. Patterson continued filming the creature as it disappeared and reappeared between trees. The adjacent diagram created by Bob Titmus shows the entire scene.[3]

(NOTE: The filming stops at about the blue dot. The continued passage of the creature shown was determined later by Titmus and is explained below.)

Gimlin wanted to immediately continue pursuit on horseback and proceeded to do so. Patterson, however, did not have his horse or his rifle and did not want to be left alone. He therefore shouted at Gimlin to return, which he did. After Patterson's situation was rectified and the camera reloaded with a second film roll, the men then followed the path taken by the creature. It was later learned (determined by Bob Titmus) that the creature had ascended a hill, sat down and probably observed the men for a time before they started to follow its trail. The men found scuffmarks in the gravel and in the creek bed which may have indicated the creature ran when it was out of the men's sight. They continued up the creek for a considerable distance and observed a rock with a wet half footprint on the surface. From that point the path led up into the mountains.

---

2 It is reasoned that either Patterson's camera noise or the fact that Patterson and Gimlin were now seen on foot made the creature uneasy. This latter theory is covered in detail later under THE FILM SITE MODEL discussion.
3 Titmus shows "Old Road Paralleling Creek." This was not an "old" road in the sense of being unused. It was built for logging operations currently in progress.

The men then returned to the film site and examined the path the creature had taken along the sandbar. They observed and filmed the creature's footprints in the soil and later made plaster casts of the left and right foot. In that part of Bluff Creek, there is a sandy clay soil with a blue-gray tinge. This type of soil holds footprints remarkably well for a long period of time. The footprints measured about 14.5-inches/36.8m long by 6-inches/15.2cm wide. Gimlin jumped off a log to see how far his footprints would sink into the soil in comparison with the creature's prints. The results were that the creature's footprints were deeper. Patterson also took movie footage of this experiment together with footage of horse prints alongside the creature's prints. Gimlin filmed Patterson making casts and also displaying the finished casts as seen here.

That same afternoon the men drove to Arcata (or Eureka) to ship the films for processing. During this time Patterson telephoned a reporter at *The Times-Standard* newspaper and related the story of the filming. The following article (exactly reprinted) was featured on the front page of the paper the next day, October 21, 1967.

A Daily Newspaper for Northwestern California and Southern Oregon

EUREKA, CALIFORNIA, SATURDAY, OCT. 21 1967   Price Per Copy: Daily 1

Mrs. Bigfoot Is Filmed!

A YAKIMA, WASH. Man and his Indian tracking aide come out of the wilds of northern Humboldt County yesterday to breathlessly report that they had seen and taken motion pictures of "a giant hominoid creature."

In colloquial words - they have seen Bigfoot!" Thus, the long sought answer to the validity and reality of the stories about the makers of the unusually large tracks lie in the some 20 to 30 feet of colored film taken by a man who has been eight years himself seeking the answer.

And as Roger Patterson spoke to The Times-Standard last night, his film was already on its way by plane to his home town for processing while he was beside himself relating the chain of events.

Patterson, 34, has been eight years on the project. Last year he wrote a book, "Do Abominable Snowmen of America Really Exist?" This year he has been taking films of tracks and other evidence all over the Northwestern United States and Canada for a documentary.

WHO WROTE THE TIMES STANDARD ARTICLE?

We do not know the name of the newspaper reporter Patterson contacted. Unfortunately, whoever it was did not show his or her name in the article (i.e., the "by" line). We are told by *The Times-Standard* that it was not uncommon in the 1960s for articles not to the show a reporter's name. *The Times-Standard* does not think it would be able to find the person's name now. Researchers are hopeful that the reporter will one day come forward or other *Times-Standard* employees will recall who he or she was and provide this information.

He has over 50 tapes of interviews with persons who have reported these findings, and including talks with two or three persons who have reported seeing these giant creatures.

BOB GIMLIN, 36, and a quarter Apache Indian and also of Yakima, has been associated with Patterson for a year. Patterson has visited the area before and last month received word of the latest discovery of the giant footprints which have become legend.

Last Saturday they arrived to look for the tracks themselves and to take some films of these, riding over the mountainous terrain on horseback by day and motoring over the roads and trails by night.

Yesterday they were in the Bluff Creek area, some 65 to 70 miles north of Willow Creek, where Notice Creek comes into it. They were some two miles into a canyon where it begins to flare out.

Patterson was still an excited man some eight hours after his experience. His words came cascading out between gasps. He still couldn't believe what he had seen, but he is convinced he has now seen a "Bigfoot" himself and he's the only man he's heard of who has taken pictures of the creature. Here is what he reported:

IT WAS about 1:30 p.m., the daylight was good, when he and Gimlin were riding their horses over a sand bar where they had been just two days before. They had both just come around a bend when "I guess we both saw it at the same time."

"I yelled 'Bob Lookit' and there about 80 or 90 feet in front of us this giant humanoid creature stood up. My horse reared and fell, completely flattening a stirrup with my foot caught in it.

"My foot hurt but I couldn't think about it because I was jumping up and grabbing the reins to try to control the horse. I saw my camera in the saddle bag and grabbed it out, but I finally couldn't control the horse anymore and had to let him go."

GIMLIN was astride an older horse which is generally trail-wise, but it too rared (sic) and had to be released, running off to join their pack horse which had broken during the initial moments of the sighting.

Patterson said the creature stood upright the entire time, reaching a height of about six and a half to seven feet and an estimated weight of between 350 and 400 pounds.

"I moved to take the pictures and told Bob to cover me. My gun was still in the scabbard. I'd grabbed the camera instead. Besides, we'd made a pact not to kill one if we saw one unless we had to."

Patterson said the creatures'(sic) head was much like a human's though considerably more slanted and with a large forehead and broad, wide nostrils.

"It's (sic) arms hung almost to its knees and when it walked, the arms swung at its sides."

PATTERSON said he is very much certain the creature was female "because when it turned towards us for a moment, I could see its breasts hanging down and they flopped when it moved."

The creature had what he described as silvery brown hair all over its body except on its face around the nose and cheeks. The hair was two to four inches long and of a light tint on top with a deeper color underneath.

"She never made a sound. She wasn't hostile to us, but we don't think she was afraid of us either. She acted like she didn't want anything to do with us if she could avoid it."

Patterson said the creature had an ambling gait as it made off over the some 200 yards he had it in sight. He said he lost sight of the creature, but Gimlin caught a brief glimpse of it afterward.

"But she stunk, like did you ever let in a dog out of the rain and he smelled like he'd been rolling in something dead. Her odor didn't last long where she'd been."

LATE LAST NIGHT Patterson was anxious to return to the campsite where they had left their horses. He had been to Eureka in the afternoon to airmail his film to partner Al De Atley in Yakima. De Atley has helped finance Patterson's expeditions.

He and Gimlin were equally anxious to return to the primitive area. "It's right in the middle of the primitive area" for the chance to get another view and more film of the creature.

He said there's strong belief that a family of these creatures may be in the area since footprints of 17, 15 and nine inches have been reported found.

The writer jested that these sizes put him in mind of The Three Bears.

"This was no bear," Patterson said. "We have seen a lot of bears in our travels. We have seen some bears on this trip. This definitely was no bear."

Patterson is also anxious today to telephone his experience to a museum administrator who is also extremely interested in the project. "He may want to bring down some dogs. We don't have dogs here."

He's not sure how much longer they will remain in the area. "It all depends."

THE IMPORTANCE OF *THE TIMES-STANDARD* ARTICLE

Neither Patterson nor Gimlin kept any record or notes of their experience at Bluff Creek. However, Patterson's contact with *The Times-Standard* reporter resulted in a "diary" of the day's events. As the contact was made very soon after the filming (at about 9:30 p.m. the same day), we can be certain everything was still very fresh in Patterson's mind.

There is, however, one discrepancy in the article that needs explaining. The reporter places the time of the men's arrival at Bluff Creek as "last Saturday." As the article was written on October 20, 1967, the "last Saturday" was October 14, 1967. We know that the two men had been in the area for about three weeks, so the reporter's arrival time is obviously incorrect.

1967						
OCTOBER						
Su	Mo	Tu	We	Th	Fr	Sa
1	2	3	4	5	6	7
8	9	10	11	12	13	14
15	16	17	18	19	20	21
22	23	24	25	26	27	28
29	30	31				

Frame 352 of the film shows the creature as it turned to look at Patterson and Gimlin.

Patterson is seen here making a cast of a footprint left by the creature.

Photo of Patterson taken in 1970.

Actual 16mm film size.

The film itself is Kodachrome II, 16mm movie film. The illustration seen here shows five film frames actual size. The sasquatch creature is about 1.2mm high in each frame. It is just visible to the naked eye. Even when the film is projected onto a screen, the creature is not very large and few details are visible. Regular size photographs from the frames have been produced and although somewhat fuzzy, give us a much better idea of what the creature looked like than the projected images.

Bob Gimlin, seen here in 1967, has never wavered in his testimony that what he saw at Bluff Creek was an actual sasquatch creature. He did not gain financially from the proceeds of the film.

Bob Gimlin (left) and Roger Patterson (right). Bob Gimlin is comparing his foot to a 16-inch /40.64cm cast of a print found by Pat Graves in October 1964 near Laird Meadow Road, Bluff Creek, California area. Gimlin is holding one of the film site casts as is Patterson.

Bob Gimlin, September 2003. I had the pleasure of meeting Bob during this month and spent considerable time with him. There is absolutely no doubt in my mind as to his testimony.

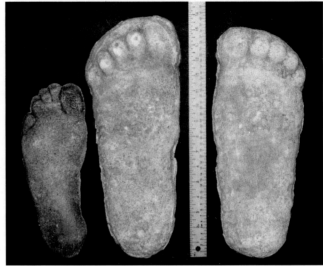

Shown here are casts of the creature's footprints along with the cast of a human foot (11.5-inches/29.2cm long) for comparison purposes. The creature's actual footprints at the film site measured about 14.5-inches/36.8cm long and 6-inches/15.2cm wide (widest part). The creature casts differ in appearance because of soil and movement conditions. (NOTE: Creature casts illustrated are copies and are therefore slightly larger than the original casts.)

All of the photographs shown above are of footprints left by the creature in the Patterson/Gimlin film. The first photograph is from the second film roll used by Roger Patterson. Patterson gave about 10-feet/3.1m of this roll to René Dahinden and parts of the roll were used in early television documentaries. Unfortunately, the actual film roll is now missing. The next three photographs were taken by Robert Lyle Laverty one day after the filming (October 21, 1967). The prints were impressed in the soil up to a depth of 1-inch/2.5cm.

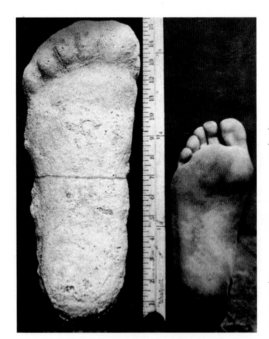

This widely publicized photograph shows René Dahinden's foot compared to a cast of the creature's foot. The comparison is not quite "fair" because plaster casts are always slightly larger than the actual foot that made the print used for the cast. Also, it is seen that René did not have very large feet. Nevertheless, we get a good sense of comparison both in size and configuration. The fact that the cast is cracked at the center indicates to me that it was used by Dahinden to make copies – such cracks are a common occurance in this process.

THE FIRST FILM SCREENING TO SCIENTISTS

Prior to the Patterson/Gimlin film, the British Columbia, Canada, government had expressed some interest in the sasquatch issue. Don Abbott, a professional with the British Columbia Provincial Museum had, in fact, gone to California in August 1967 (at the request of John Green and René Dahinden) and personally inspected sasquatch footprints. As a result of Abbott's involvement, Patterson agreed to screen the film to scientists at the University of British Columbia on October 26, 1967. Those people who attended this screening were: Roger Patterson, Robert Gimlin, John Green, René Dahinden, Robert Titmus, David Hancock, Dr. Ian McTagert Cowan, Dr. Beverly Green (both from UBC), Frank Beebe, Don Abbott, Charles Guiguet (all from the Provincial Museum). While we do not have any "official" statements from the university scientists, John Green provides the following account and analysis of the scientific views that were expressed.

Scientists among the group who watched the first public screening of the Patterson-Gimlin film in 1967 raised three negative comments about the creature in the film that continue to surface occasionally, despite the fact that two of them are totally wrong and the third is very questionable.

The first is that the creature has female breasts yet it walks like a man.

The second is that the creature has a sagittal crest, which is characteristic only of male gorillas.

The third is that no higher primate has breasts that are covered with hair.

In fact it would have been an indication of a hoax if the creature did not walk "like a man." Among higher primates only the human female has a walk different from that of the male, because wider pelvises are needed to give birth to human infants, which have exceptionally large heads.

As to the sagittal crest, its function is to provide an anchorage for large jaw muscles, and it is related to size, not to sex. Since the creature in the film is bigger than a male gorilla, lack of a sagittal crest would have been an indication of a hoax.

And female apes do have some hair on their breasts, even though they live only in hot climates. Apes adapted to climates as far north as the Bering Strait land bridge could surely be expected to have a great deal more hair.

None of the scientists at that first screening was a primatologist or physical anthropologist, and their opinions were asked for after a brief look at the film with no time for research or reflection, so their comments, while damaging, are probably excusable. There is no such excuse for their modern colleagues.

Here is what Dr. Grover Krantz, who was a physical anthropologist, had to say in his book ***Bigfoot Sasquatch Evidence***:

"Of course the female sasquatch walks more like a man than a woman, and that is exactly how she should walk."

"Human females generally walk rather differently from males, but there is no such contrast in apes. In our species the female pelvis is relatively much wider at the level of the hip sockets than is the male pelvis. This results from the very large birth canal that is required for our large-headed newborns. Apes are born with much smaller brains and their two sexes have more nearly the same pelvic design.

"Of course the female sasquatch walks more like a man than a woman, and that is exactly how she should walk. (p.116-117)

".....a sagittal crest is not a male characteristic...on the contrary it is a consequence of absolute size alone. As body size increases, brain size increases at a slower rate than does the jaw, so a discrepancy develops between these two structures. When jaw muscles become too large to find sufficient attachment on the side of the braincase a sagittal crest develops.

"That size threshold is regularly crossed by all male gorillas and a few females, by most male orang utans and by no other known primates. The evident size of the sasquatch easily puts their females well over that threshold, and a sagittal crest would be an automatic development. (p.304-305)

"That the species should have enlarged breasts at all (a human trait) is also a point of contention to some critics....But that they would be hair covered in a temperate climate seems perfectly reasonable to me." (p. 119)

Roger Patterson died of Hodgkin's disease in 1972. He was 39 years old. Robert Gimlin (b. 1931) still lives in Yakima County, Washington State. Over the last 37 years the film has been subjected to intense examination and cannot be proven to be a fabrication. Findings indicate that the creature filmed was a natural creature.

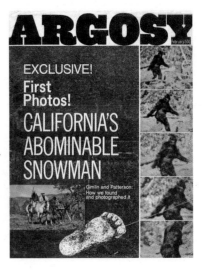

The first major magazine article on Patterson and Gimlin's experience was published by Argosy *magazine in February 1968. The magazine cover is shown here. The article was written by Ivan T. Sanderson (previously mentioned) who had ironically influenced Patterson to take up sasquatch research. In the photograph seen of Patterson and Gimlin on horseback, Gimlin is dressed as an Indian scout. Partly of First Nations heritage, Gimlin dressed this way for Appaloosa shows in which he frequently participated. The photograph was taken by Patricia Patterson for personal purposes before the Bluff Creek event. It was apparently selected by* Argosy *for effect.*

PHOTOGRAPHS FROM THE PATTERSON/GIMLIN FILM

The twelve photographs shown here are considered the clearest images of the creature seen in the film. I present reduced retakes of the Cibachrome prints first. The Cibachromes are closeup and cropped prints 3.5-in. x 4.5-in. (8.9cm x 11.4cm) made in the early 1980s. Next I provide enlargements of the creature at about 80 times the size of the image in the actual film frames. Lastly, I provide the full frame photographs. The actual film frame numbers are shown on all photographs.

NOTES

1. The film contains 953 film frames. Many frames contain little or no information because Patterson was running while filming and at one point stumbled. Numerous frames, however, show reasonably clear images of the creature. Nevertheless, as it is a movie film, there are only marginal differences between frames either side of the frame numbers presented. For example, there is very little difference between frames 351, 352 and 353. Frame 352 was simply judged to be the best by two people.

2. The only credible details that can be seen on the enlargements are those that can be seen with the naked eye. In other words, more detail or additional details seen with a magnifying glass or on a further enlargement do not have credibility.

Enlargements of the Creature - Patterson/Gimlin Film

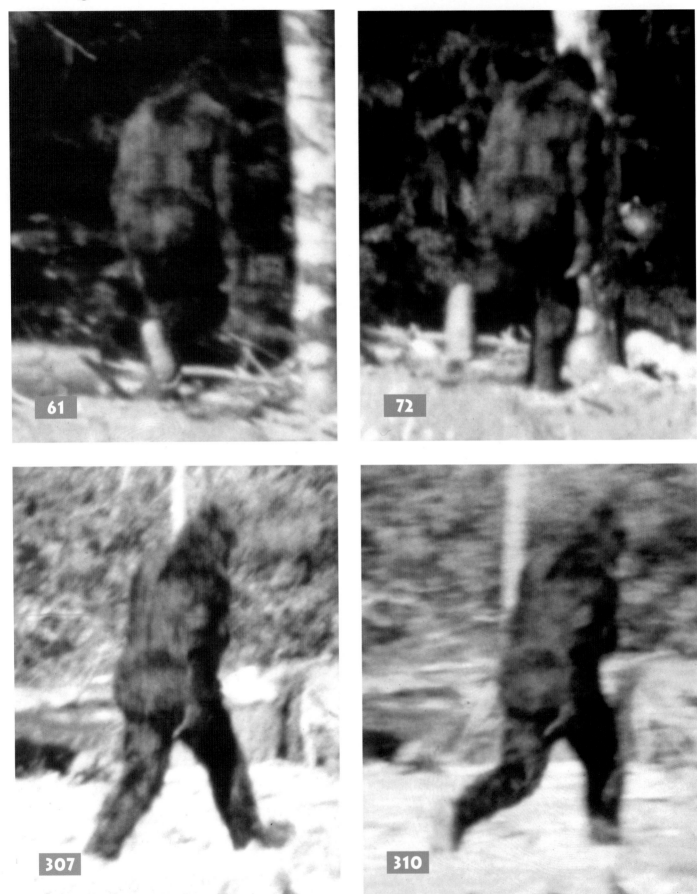

Enlargements of the Creature - Patterson/Gimlin Film

Enlargements of the Creature - Patterson/Gimlin Film

Naturally, Roger Patterson did not see the creature as we see it in the foregoing photographs. To experience what he saw, one needs to look at the full frames, which are now presented. Please note, however, that I am missing three (3) frames in the series. Also, the photos have been cropped slightly for this presentation.

Frame 352 has a different color texture because what is shown here came from a different print. The scene in this frame was highly popular. It was enlarged to postcard size and provided to authors and publishers.

AERIAL VIEW OF THE FILM SITE

The following is an aerial view of the film site taken in 1971. A person is shown walking the path taken by the creature. The approximate position of Patterson or the camera is shown with a red box. The creature's path is shown with a red arrow. We can see, therefore, that the photograph was taken from an elevated position far to the right of Patterson's position or the camera position at the time the movie film was taken. The photograph is less two trees present at the filming that had fallen down prior to the time this photograph was taken. The tree marked with a red "X" is the second or center tree, seen in the film frames.

PATTERSON'S CAMERA AND FILMING SPEED

The adjacent Kodak ad shows the camera Roger Patterson used to film the creature at Bluff Creek.

Considerable controversy resulted over the speed of the actual filming (frames per second – FPS). Unfortunately, Patterson did not remember at which speed the dial was set when he filmed the creature. The camera has five settings: 16, 24, 32, 48, and 64 FPS. It has been established that the camera was set at either 16 FPS or 24 FPS. According to Dr. D. W. Grieve[1], if the camera was set at 16 FPS, the creature's gait (walking pattern) is quite unlike a human being's gait. At 24 FPS, the gait cannot be distinguished from a normal human gate. Considerable research was performed on the filming speed question by the Russian hominologist, Igor Bourtsev. Patterson took some movie footage (initial shots) while running. The jerking and shaking of his movements are reflected in the film. Bourtsev took the vacillation of images on the film and related them with Patterson's steps and movements. He concluded that if the camera was set at 24 FPS Patterson had to be moving at a rate of six steps per second which is physically impossible. Patterson's maximum rate would be four steps per second and this rate corresponds with a filming speed of 16 FPS. Given Bourtsev's conclusion is correct, then the creature filmed was probably a natural sasquatch being.

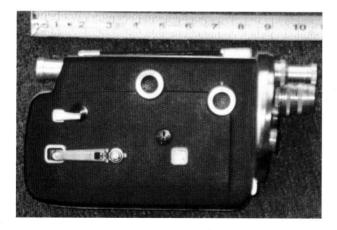
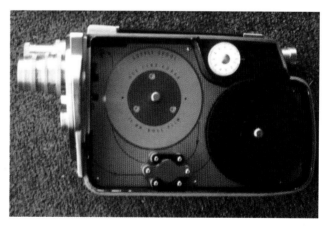

The actual camera shown here (second photo shows camera opened) is similar to Patterson's camera - the only difference being that this camera is equipped with a rotating turret.

1 See Authoritative Conclusion on the Patterson/Gimlin Film for the complete report by Dr. Grieve.

THE FILM SITE MODEL

The photograph shown here is of a model I have constructed of the film site at Bluff Creek. The model scale is about 1-inch equals 9-feet. The entire model is 21-inches/53.3cm long by 16.25-inches/41.3cm wide. Actual measurements as they relate to objects seen are presented later. To my knowledge, no one has previously attempted to build a model of this nature. Indeed, I would not have attempted it myself were it not for a "strange" document given to me some five or six years ago by the late René Dahinden. The document, a photocopy of film frame No. 352, with numerous lines and symbols (seen below), was given to me without explanation. I often wondered what it was all about, but just never got around to discussing it with René. René had performed a lot of measurements at the film site in the early 1970s. He drew a diagram (not to scale) of the site that was published by Daniel Perez in his book, *Bigfoot Times*, 1992. René showed little circles representing trees and stumps, but what trees and what stumps? By the time I started analyzing the diagram, René had passed away, and I did not know of anyone else who would have the answers. René identified some of the objects in his sketch with Roman numerals. He had also marked various objects in photographs of the site (which he had given me) with Roman numerals. When I matched the photographs to the diagram, I started to see how the site "came together." However, the photographs showed only a few objects seen in the diagram, so I could not associate other objects.

From "out of nowhere" the thought came to me to get the "strange" document previously mentioned. As soon as I laid it on my desk, the whole film site unfolded before my eyes. This document was the "key" to the diagram. The Roman numerals and other symbols matched those on the diagram so that one could see the exact placement of most of the important objects in the film frames. I started "mapping out" the site and before long was thoroughly engrossed in constructing a proper model.

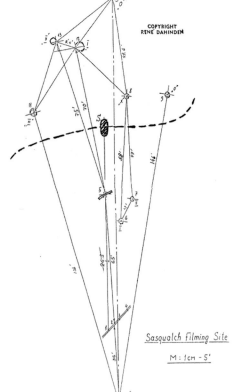

Unfortunately, René had omitted to provide any meaningful horizontal measurements. While some of these measurements could be reasonably calculated, others had to be estimated using actual film frame photographs, other photographs (including the aerial shot of the site) and old-fashioned logic. In some areas, I was able to get missing information from a rough site sketch made by John Green (also published by Dan Perez). I visited John with my first model "draft" as it were, and he showed me film footage he took of the site in 1968. I am confident the model is very close to the original scene.

"I am confident the model is very close to the original scene."

In talking to John about events in 1967, he and the other sasquatch researchers felt it would only be a matter of months before a sasquatch creature would be "brought in." For this reason, we do not have highly detailed information on the film site. Nevertheless, what we have is sufficient and we can thank René and John for the great work they did in this regard.

There is an old saying that "cameras do not lie." While this saying is true, cameras do something just as bad, they deceive. While objects seen in photographs are exactly as they are in real life, their relationship to other objects is a totally different matter. In the Patterson/Gimlin film frames we commonly see, we are led to believe that the creature is within a few feet of individual trees, stumps, logs and the forest in the background. This conception is totally incorrect as can be seen in the foregoing model photograph.

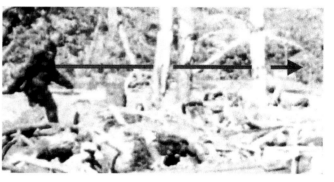

The most evident deception is the position of the three trees directly in front of the creature in frame 352 together with the forest debris in the frame foreground. In the film, it is seen that the creature goes behind the first tree, in front of the second tree, and then behind the third tree.

The first tree is about 48-feet/14.6m away, directly towards the camera. The second tree is about 10-feet/3.1m further back from the creature's path. The third tree is about 39-feet/11.9m away, again directly towards the camera. The forest debris begins at over 30-feet/9.1m from the creature's path. The first and third trees are in this debris. Roger Patterson was just 5-feet, 2-inches/1.6m tall and we believe he crouched down, or was perhaps still on his knees after stumbling when these frames were filmed. As a result of the low camera height, forest debris in the foreground concealed the space between the debris and the creature.

"As a result of the low camera height, forest debris in the foreground concealed the space between the debris and the creature."

The following illustrations show the relative position of objects seen in the film frames and associated measurements.

Relative Positions of Objects Seen in the Film Frames

Frame 352

Site Model

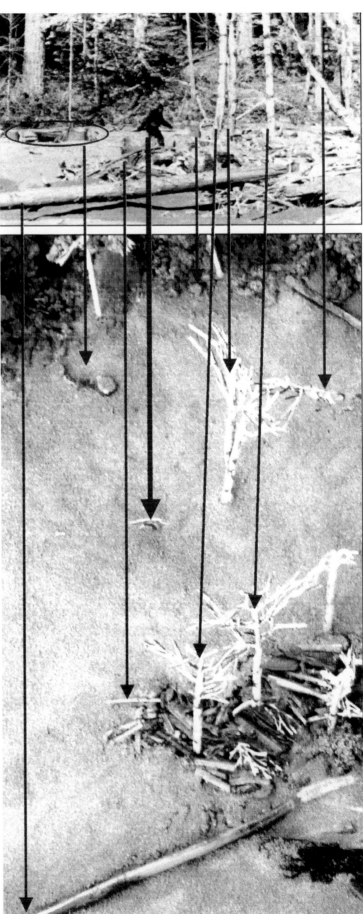

Bluff Creek Film Site Measurements - External Points

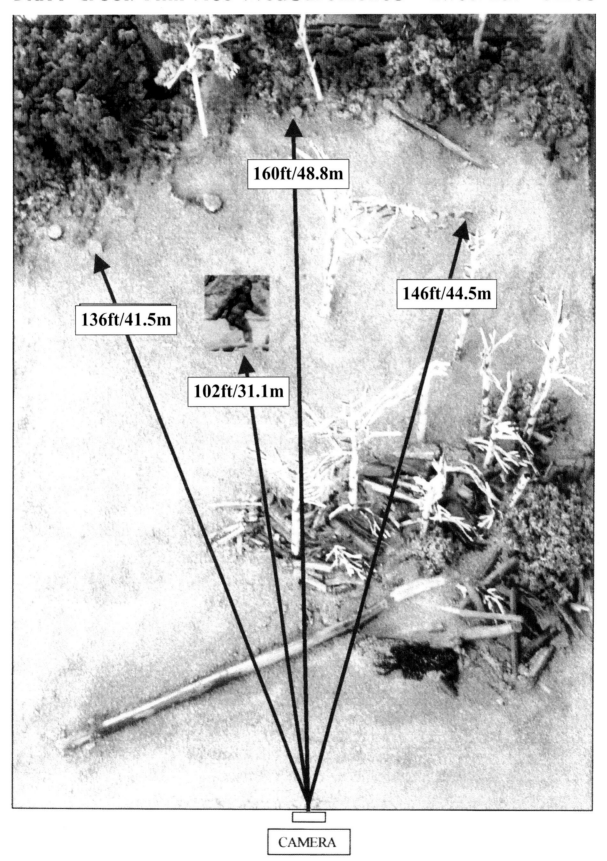

Bluff Creek Film Site Measurements - Internal Points

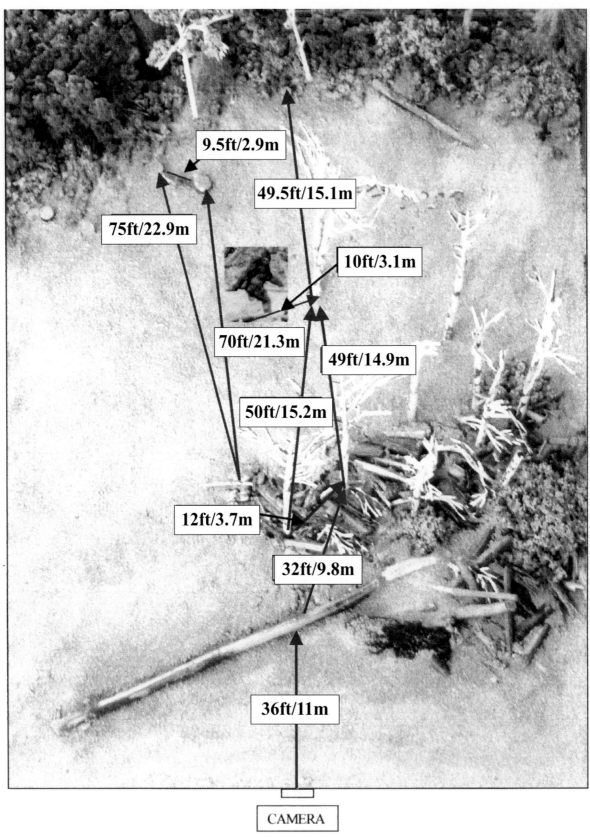

It is highly interesting to view the model from different angles. When seen from the left, we can get an idea of what the creature saw as it crossed the gravel sandbar.

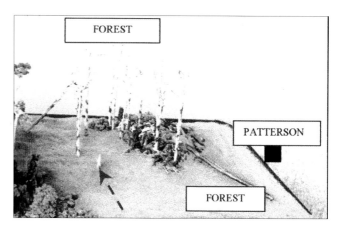

We are reasonable sure that the creature was either not overly concerned with Patterson and Gimlin's presence or in fact aware of their presence until it reached about the position shown in the model. Here, it turns and looks directly at the men. At this point the forest to its left has a steep mountains side. The forest behind (from where it came) is about an equal distance to the forest directly ahead. These conditions likely account for the fact that the creature did not dart into the forest to its left or turn back – a contentious point with some people. Further, the creature possibly took into consideration the trees and debris to its right - on the little island as it were. This "cover" effectively blocked a clear view from Patterson's position. Indeed, when we view the film beyond the model boundary, we can see how difficult it would have been to get a clear rifle sighting on the creature. All of these deductions, of course, are pure speculation. However, they do justify in my mind why the creature chose to just keep moving ahead. Do I think the creature was frightened? I think it was terrified.

The expression on its face in the last clear film frame (frame 364, seen here) indicates to me that it was both terrified and a little confused as to what to do. Under the circumstances, it did what we are told to do when we are confronted with a wild animal - calmly and quickly, but don't run, put distance between yourself and the animal.

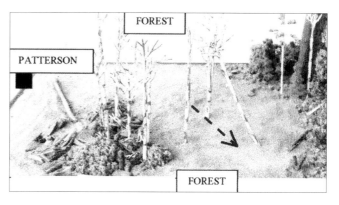

Viewed from the right, the model illustrates the importance of the little "island" to the creature. Keep in mind that the island is a tangled mass of forest material. One could not run through it so it definitely put something between the creature and the intruders. The fact that Patterson and Gimlin just stayed where they were (did not pursue the creature) also probably resulted in the creature continuing at a steady pace. Had either or both men rushed forward about 100-feet/30.5m, I think the creature would have been out of sight very quickly.

Viewed from the back, we can see most clearly how Patterson and Gimlin had an unobstructed view of the creature for a considerable distance. By the same token, the creature had an unobstructed view of the men, but as mentioned, did not show concern until it reached the position shown. We might reason that the creature was not overly concerned with the men until it came within earshot of the clicking made by Patterson's camera and then saw something being pointed at it (i.e., it then realized it was the "target" of some action).

Another highly possible theory offered to me by Dave Hancock is that it was at this point the creature saw Patterson as a man rather than a horse. Here I must explain things a little. When the creature first saw Patterson and Gimlin they were on horseback. The creature looked up and saw three horses, all of them reacting to its presence. Horses were not a threat to the creature and it knew this. It simply walked away. Patterson then ran on foot in a parallel course to the creature and stopped. Gimlin rode towards the creature on horseback and although Gimlin believes the creature saw him, it did not react. However, when the creature saw Patterson (and by this time also Gimlin) on foot, it then realized it was being observed by men and became alarmed.

John Green (left) and author at John's place after discussion of the film site model.

I believe one of the most important points presented in this whole discussion is that the forest behind the creature (i.e., to its left) is more than 160-feet/48.8m from the camera. I do not know its exact location, however, it is beyond the furthest object measured by Dahinden. If one looks at photographs of the film frames and discovers an unusual "shape" in the forest, it must be taken into account that the "shape" is over 58-feet/17.7m further back than the creature seen. We can only "just see" general fuzzy details on the creature at 102-feet/31.1m. At 160-feet/48.8m we would only be able to distinguish its overall appearance which of course would be much smaller in size.

The film site is certainly no longer like the model. Even as early as 1972 at least three of the four trees used for measurements had fallen down. Over the last 37 years, the forest has evidently reclaimed most of the clearing. On visiting what is claimed to be the film site in September 2003, I could not recognize any of the trees nor find the large log and stumps seen in the model. It appears that at some point, Bluff Creek flooded and eroded a large section to the right (facing) side of the site. The creek then appears to have returned back to its original width and course, leaving a rocky river bed in what was once part of the clearing. However, while I am sure the location to which I was taken was the general film site area, I am not sure it was the actual spot we see in the film site model.

Bob Gimlin (left) and author with the film site model at Willow Creek, California, September 2003.

DIMENSIONS OF THE CREATURE SEEN IN THE PATTERSON/GIMLIN FILM

The dimensions shown here were calculated by John Green. John based his calculations on the creature's foot size (14.5-inches). He verified the height he established by photographing a person at the film site from the same distance as the creature is seen in the film, and then registering both images. John points out: *"that since the creature seen in the film is blurry and it has hair or fur of unknown thickness, the measurements are not exact. Measurements from the back are considerably less exact than from the side, as this view of the creature is at a much greater distance."*

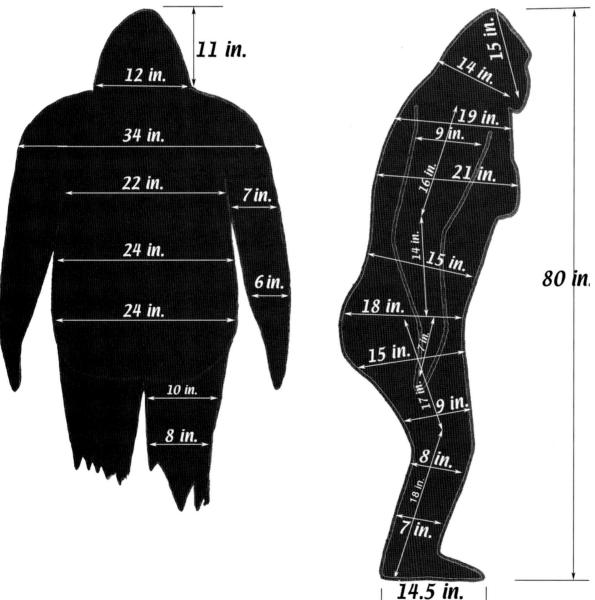

THE CREATURES WALKING PATTERN

Many, if not most sasquatch footprints indicate that the creatures have a straight walking pattern. In other words, there is minimal or nonexistent "angle of gait" (e.g., the creature walks as though it were on a tightrope). The creature in the Patterson/Gimlin film definitely had a straight walking pattern. This fact, however, was not confirmed until recently (2003) to my knowledge. (NOTE: I show examples of four different sets of consecutive footprints later in the **Footprint and Cast Album**.)

This illustration shows the Patterson/Gimlin film creature's footprints in a series. The illustration was created by Yvon Leclerc who took the individual film frames and registered them to provide a single picture. The red line indicates the creature's path which sort of ambles to the right. If we take this illustration and draw a straight line from the first to the last prints, it can be seen that the line comfortably hits each print.

Performing this little test on regular human footprints would not yield the same results. The line would completely miss every other print. I say "regular" human footprints because humans can definitely walk in a straight line. Indeed, female fashion models are taught to walk in this way because it is far more pleasing to the eye in that particular business.

In this photograph of Patterson casting a footprint at the film site, we would expect that the next footprint and subsequent prints should be seen in the distance. However, it is reasoned that the other prints are behind Patterson as indicated by the yellow line.

Here I explore one of the Patterson/Gimlin film frames (frame 323) for possible evidence of the creature's footprints. I have identified marks that might be footprints. The third impression appears to show toes so I am reasonably sure it is such.

Given the depth of foot impressions made by the creature in the film, it is a little odd that we do not see any highly definite impressions in the film frames rather than these somewhat speculative and marginal impressions. Nevertheless, the marks I have identified surprisingly appear to line up with the prints previously shown in Yvon Leclerc's footprint registration.

There is, however, one point that must be addressed. We know that the creature either stepped on, or stepped very close to, the wood fragment identified. The photograph on the right shows the moment of contact or near contact. This fact indicates that there should be a footprint impression very close to the fragment. However, such is not the case. The third mark, which is the obvious associated mark, appears to be too far away from the fragment. The answer here could be that the wood fragment moved slightly after being stepped on (sort of spun forward towards the camera a few inches).

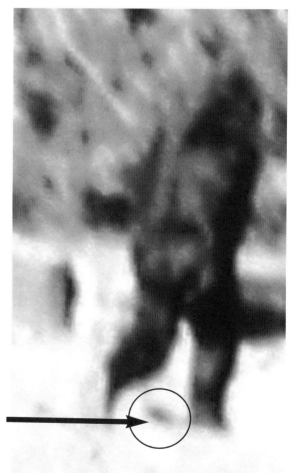

THE WOOD FRAGMENT

Frame 352

René Dahinden visited the Patterson/Gimlin film site in 1971 (about four years after the filming) and observed what he considered to be the same wood fragment in exactly the same spot as seen in the film frames (identified detail on adjacent film frame photograph). The fragment was either stepped on, or nearly stepped on, by the creature as it headed along the sandbars as previously discussed. Dahinden took the fragment with him when he returned to his home. The following are photographs of the fragment Dahinden retrieved.

This photograph shows the probable position or orientation of the fragment in the film frames.

In this photograph, the fragment has been positioned for measurement. The total length including extremities is about 26.25-inches/66.8cm.

It was reasoned that if the fragment retrieved by Dahinden was in fact the fragment seen in the film frames, then it might be used to calculate the height of the creature.

The fact that the wood fragment had to remain in the same spot for some four years greatly detracts from the possibility that it is the fragment seen in the film frames. Unfortunately, photographs taken of the film site at the time (1971) are too difficult to reconcile with the film frames. The following is one such photograph.

René Dahinden's son, Erik, is seen here at the film site standing in the path taken by the creature as determined by René. The wood fragment is circled. René himself drew the circle.

The fuzzy image of the fragment in the film frames makes it very difficult to do an accurate calculation. Further, we do not know the camera angle.

I first learned that René had retrieved the fragment (or a fragment) from the film site while watching the film with him. He pointed to it and stated, " I have that piece of wood." He told me he had carefully compared the film frames to the site and noticed that the fragment was still there. Later, when he showed me the fragment in conjunction with possible calculations, I asked why he had retrieved it. He laughed and said, "Seeing she probably stepped on it, I thought I could get some "vibes." I performed rough calculations at the time not knowing the time frame (I thought he had obtained the fragment very soon after the filming). My calculation supported an established creature height of about 87-inches. When the time frame was brought to my attention, I naturally had to agree there was a good possibility that it was not the actual fragment seen in the film. Nevertheless, René remained steadfast that it was the same fragment. He then told me that he had used the physical fragment to calculate the distance of the creature from the camera. In time, he gave me a strip of the film to see if I could confirm that the creature did in fact step on, or nearly step on, the fragment and such definitely is the case (I believe it actually stepped on it).

After René had passed away and I stumbled on the key to his diagrams, (as discussed under **The Film Site Model**), I was surprised to see that he had measured a camera distance of 101-feet/30.8m, only one foot/30.5cm short of Dr. Krantz's estimate of 102-feet/31.1m which was arrived at under a different process. I must admit that this finding made me wonder. If it is not the same fragment, then it is highly coincidental that another fragment of reasonable similarity was at the distance René established.

ARTISTIC IMAGES OF THE CREATURE SEEN IN THE PATTERSON/GIMLIN FILM

Given a reasonable portion of imagination, your author performed the following study in 1996 on the creature's head as seen in frame 352. The original image used (top left photograph) was derived from a blowup of frame 352. The enhancements were done with pastels on a laser color copy. I closed the creature's mouth to give it a more natural look.

The final picture met with remarkable acceptance. One sasquatch eyewitness at a symposium stated most positively, "That was the creature I saw."

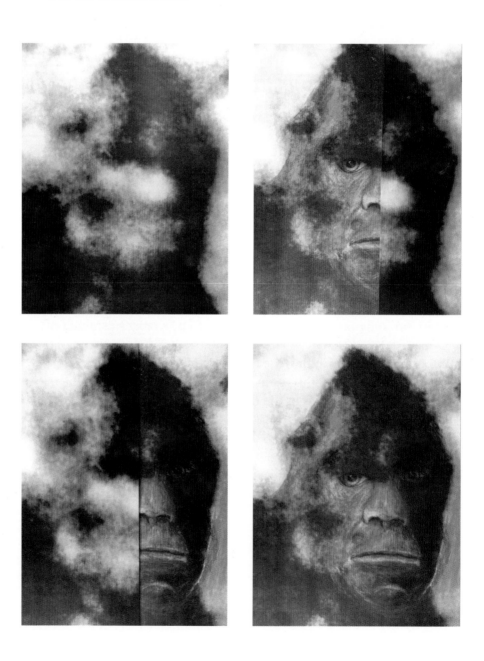

Much later, in working with Yvon Leclerc, I provided him with the images on the left (frames 339 and 350, both derived from actual film frames) and he produced the computer enhancements shown on the right.

Frame 339

Frame 350

Penny Birnam Sculptures

These remarkable sasquatch head sculptures were created by Penny Birnam specifically for an exhibit at the Vancouver Museum. All heads are based on the creature seen in the Patterson/Gimlin film. Penny, seen here with your author, is a Vancouver sculptor with a fervent interest in conservation. She graduated from Emily Carr College of Art in 1986 and has been earning her living by creating sculptures of endangered species and other animals since that time. In Penny's own words, "The animals are my sympathetic magic. Like voodoo dolls in reverse, they carry my love and respect into the world, and my hope that their new owners will become protective of them."

"The Patterson-Gimlin film is an authentic documentary of a genuine female hominoid, popularly known as Sasquatch or Bigfoot, filmed in the Bluff Creek area of northern California not later than October 1967."

Dmitri Bayanov

Igor Bourtsev

AUTHORITATIVE CONCLUSIONS ON THE PATTERSON/GIMLIN FILM

Over the last thirty seven years, the Patterson/Gimlin film has undergone rigorous examination by highly professional and dedicated people. The following are the conclusions reached by the main and most prominent people.

Conclusions reached by the Russian hominologists Dmitri Bayanov and Igor Bourtsev: (The following is a reprint from the book *America's Bigfoot: Fact Not Fiction*, by Dmitri Bayanov (Crypto Logos, Moscow, Russia, 1997).

We have subjected the film to a systematic and multifaceted analysis, both in its technical and biological aspects. We have matched the evidence of the film against the other categories of evidence and have tested the subject with our three criteria of distinctiveness, consistency and naturalness. The film has passed all our tests and scrutinizes. This gives us ground to ask: Who other than God or natural selection is sufficiently conversant with anatomy and biomechanics to "design" a body which is so perfectly harmonious in terms of structure and function[1]

The Patterson-Gimlin film is an authentic documentary of a genuine female hominoid, popularly known as Sasquatch or Bigfoot, filmed in the Bluff Creek area of northern California not later than October 1967.

Until October 1967, we had lots of information on relict hominoids but they remained inaccessible to the investigators sense of vision. We were dealing then with the underwater part of the "iceberg," as it were. October 1967 was the time when the fog cleared and the tip of the iceberg came into view. True, we still can't touch or smell this "tip," and have to be content with viewing it in the film and photographs obtained from the film. But in this we are not much different from the physician who studies a patient's bones without ever meeting the particular patient - just looking at the x-rays. Or from the geologist, who studies the geology of Mars by looking at the photographs of its surface.

The difference is of course that in the geologist's case seeing is believing and, besides, he has all the might of modern science at his disposal. Those photographs cost a couple of billion dollars and nobody dares to treat them frivolously. The Sasquatch investigator, on the other hand, offered his photographic evidence to be studied by science for free and the evidence was not taken seriously.

---

[1]. I have deliberately phrased this sentence after one in Napier's book.

According to Dr. Thorington, Jr., of the Smithsonian, "...one should demand a clear demonstration that there is such a thing as Bigfoot before spending any time on the subject." If by a clear demonstration Dr. Thorington means a live Bigfoot be brought to his office, then it would be more of a sight for a layman than for the discriminating and analytical mind of a scientist.

Relict hominoid research is of special, potentially unlimited value for science and mankind. Thanks to the progress of the research, we know today that manlike bipedal primates, thought long extinct, are still walking the earth in the second half of the 20th century. We also know how such a biped looks and how it walks, this knowledge being available now to anyone who wants to use their eyes.

We are indebted for this breakthrough to the late Roger Patterson, who filmed a relict hominoid in northern California in 1967, but who, to our sorrow, was not destined to witness the full triumph of his achievement.

People readily believe photographs taken on the Moon but many do not believe the Patterson-Gimlin film taken here on Earth, showing something of incalculable value for science. They do not believe it because Patterson and his assistant, Bob Gimlin, were men with no academic authority to back their claim.

And so, René Dahinden stepped forth and traveled to Moscow with his own hard earned money to have the film analyzed and appraised in a scientific manner.

This has been done and the result is presented in this paper. The marriage of Russian theory and American practice in hominology has proven to be happy and fertile. By joining forces, we have established not only the authenticity of the film but also that the Sasquatch is part of the natural environment of North America, and its most precious part at that. May we offer this conclusion as our modest contribution to the cause of friendship and cooperation between the peoples of the Soviet Union and North America.

The search for humanity's living roots is a cause for all mankind and this makes us look forward to new international efforts in this intriguing investigation.

The success of this research is a triumph of broadmindedness over narrow-mindedness and serves as an example to the world at large which seems to be in dire need of such a lesson.

March 1977

"Thanks to the progress of the research, we know today that manlike bipedal primates, thought long extinct, are still walking the earth . . ."

"By joining forces, we have established not only the authenticity of the film but also that the Sasquatch is part of the natural environment of North America."

Conclusions Reached by Dr. Dmitri D. Donskoy, Chief of the Chair of Biomechanics at the USSR Central Institute of Physical Culture, Moscow:

*The following is reprinted from the book, **Bigfoot / Sasquatch, The Search for North America's Incredible Creature** by, Don Hunter with René Dahinden, (McClelland & Stewart Inc., 1993, originally published in 1973).*

Qualitative Biomechanical Analysis of the Walk of the Creature in the Patterson Film

As a result of repeated viewings of the walk of the two-footed creature in the Patterson film and detailed examination of the successive stills from it, one is left with the impression of a fully spontaneous and highly efficient pattern of locomotion shown therein, with all the particular movements combined in an integral whole which presents a smoothly operating and coherent system.

In all the strides the movement of the upper limbs (they can be called arms) and of the lower limbs (legs) are well coordinated. A forward swing of the right arm, for example, is accompanied by that of the left leg, which is called crosslimb coordination and is a must for man and natural for many patterns of locomotion in quadrupeds (in walking and trotting, for instance).

The strides are energetic and big, with the leg swung forward. When man extends the leg that far he walks very fast and thus overcomes by momentum the "braking effect" of the virtual prop which is provided by the leg put forward. Momentum is proportional to mass and speed, so the more massive the biped the less speed (and vice versa) is needed to overcome the braking effect of legs in striding.

The arms move in swinging motions which means the muscles are exerted at the beginning of each cycle after which they relax and the movement continues by momentum. The character of arm movements indicates that the arms are massive and the muscles strong.

After each heel strike the creature's leg bends, taking on the full weight of the body, and smoothes over the impact of the step acting as a shock-absorber. During this phase certain muscles of the legs are extended and become tense in preparation for the subsequent toe-off.

In normal human walk such considerable knee flexion as exhibited by the film creature is not observed and is practiced only in cross-country skiing. This characteristic makes one think that the creature is very heavy and its toe-off is powerful, which contributes to rapid progression.

In the swinging of the leg, considerable flexion is observed in the joints, with different parts of the limb lagging behind each other: the foot's movement is behind the shank's which is behind the hip's. This kind of movement is peculiar to massive limbs with

Dr. Dmitri D. Donskoy

well relaxed muscles. In that case, the movements of the limbs look fluid and easy, with no breaks or jerks in the extreme points of each cycle. The creature uses to great advantage the effect of muscle resilience, which is hardly used by modern man in usual conditions of life.

The gait of the creature is confident, the strides are regular, no signs of loss of balance, of wavering or any redundant movements are visible. In the two strides during which the creature makes a turn to the right, in the direction of the camera, the movement is accomplished with the turn of the torso. This reveals alertness and, possibly, a somewhat limited mobility of the head. (True, in critical situations man also turns his whole torso and not just head alone.) During the turn the creature spreads the arms widely to increase stability.

In the toe-off phase the sole of the creature's foot is visible. By human standards it is large for the height of the creature. No longitudinal arch typical of the human foot is in view. The hind part of the foot formed by the heel bone protrudes considerably back. Such proportions and anatomy facilitate the work of the muscles which make standing postures possible and increase the force of propulsion in walking. Lack of an arch may be caused by the great weight of the creature.

> "The movements are harmonious and repeated uniformly from step to step..."

The movements are harmonious and repeated uniformly from step to step, which is provided by synergy (combined operation of a whole group of muscles).

Since the creature is man-like and bipedal its walk resembles in principal the gait of modern man. But all the movements indicate that its weight is much greater, its muscles especially much stronger, and the walk swifter that that of man.

Lastly, we can note such a characteristic of the creature's walk, which defies exact description, as expressiveness of movements. In man this quality is manifest in goal-oriented sporting or labour activity, which leaves the impression of the economy and accuracy of movements. This characteristic can be noted by an experienced observer even if he does not know the specifics of given activity. "What need be done is neatly done" is another way of describing expressiveness of movements, which indicates that the motory system characterized by this quality is well adapted to the task it is called upon to perform. In other words, neat perfection is typical of those movements which through regular use have become habitual and automatic.

> "And all these factors taken together allow us to evaluate the walk of the creature as a natural movement without any signs of artfulness which would appear in intentional imitations."

On the whole, the most important thing is the consistency of all the above mentioned characteristics. They not only simply occur, but interact in many ways. And all these factors taken together allow us to evaluate the walk of the creature as a natural movement without any signs of artfulness which would appear in intentional imitations.

At the same time, with all the diversity of human gaits, such a walk as demonstrated by the creature in the film is absolutely nontypical of man.

Conclusions Reached by Dr. D. W. Grieve, reader in Biomechanics, Royal Free Hospital School of Medicine, London, England:

The following report is based on a copy of a 16mm film taken by Roger Patterson on October 20th, 1967, at Bluff Creek, Northern California which was made available to me by Rene Dahinden in December 1971. In addition to Patterson's footage, the film includes a sequence showing a human being (height 6 ft. 5 1/2 in./196.9cm) walking over the same terrain.

The main purpose in analysing the Patterson film was to establish the extent to which the creature's gait resembled or differed from human gait. The bases for comparison were measurements of stride length, time of leg swing, speed of walking and the angular movements of the lower limb, parameters that are known for man at particular speeds of walking.[1] Published data refer to humans with light footwear or none, walking on hard level ground. In part of the film the creature is seen walking at a steady speed through a clearing of level ground, and it is data from this sequence that has been used for purposes of comparison with the human pattern. Later parts of the film show an almost full posterior view, which permits some comparisons to be made between its body breadth and that of humans.

The film has several drawbacks for purposes of quantitative analysis. The unstable hand-held camera gave rise to intermittent frame blurring. Lighting conditions and the foliage in the background make it difficult to establish accurate outlines of the trunk and limbs even in unblurred frames. The subject is walking obliquely across the field of view in that part of the film in which it is most clearly visible. The feet are not sufficiently visible to make useful statements about the ankle movements. Most importantly of all, no information is available as to framing speed used.

Body shape and size

Careful matching and superposition of images of the so-called Sasquatch and human film sequences yield an estimated standing height for the subject of not more than 6 ft. 5 in/1.96m. This specimen lies therefore within the human range, although at its upper limits. Accurate measurements are impossible regarding features that fall within the body outline. Examination of several frames leads to the conclusion that the height of the hip joint, the gluteal fold and the finger tips are in similar proportions to the standing height as those found in humans. The shoulder height at the acromion appears lightly greater relative to the standing height (0.87:1) than in

Dr. Donald W. Grieve

humans (0.82:1). Both the shoulder width and the hip width appear proportionately greater in the subject creature than in man (0.34:1 instead of 0.26:1; and 0.23:1 instead of 0.19:1, respectively). If we argue that the subject has similar vertical proportions to man (ignoring the higher shoulders) and has breadths and circumferences about 25 per cent greater proportionally, then the weight is likely to be 50-60 per cent greater in the subject than in a man of the same height. The additional shoulder height and the unknown correction that should be allowed for the presence of hair will have opposite effects upon an estimate of weight. Earlier comments[2] that this specimen was just under 7 ft. in height and extremely heavy seem rather extravagant. The present analysis suggests that Sasquatch was 6 ft. 5-in (1.98m) in height, with a weight of about 280 lb (127kg.) and a foot length (mean of 4 observations) of about 13.3 in. (34cm.).

Timing of the gait

Because the framing speed is unknown, the timing of the various phases of the gait was done in terms of the numbers of frames. Five independent estimates of the complete cycle time were made from R. toe-off, L. toe-off, R. foot passing L., L. foot passing R. and L. heel strike respectively giving:

Complete cycle time = 22.5 frames (range 21.5-23.5). Four independent estimates of the swing phase, or single support phase for the contralateral limb, from toe-off to heel strike, gave: Swing phase or single support = 8.5 frames (same in each case). The above therefore indicates a total period of support of 14 frames and periods of double support (both feet on the ground) of 2.75 frames. A minimum uncertainty of ± 0.5 frames may be assumed.

Stride length

The film provides an oblique view and no clues exist that can lead to an accurate measurement of the obliquity of the direction of walk which was judged to be not less than 20° and not more than 35° to the image plane of the camera. The obliquity gives rise to an apparent grouping of left and right foot placements which could in reality have been symmetrical with respect to distance in the line of progression. The distance on the film between successive placements of the left foot was 1.20x the standing height. If an obliquity of 27° is assumed, a stride length of 1.34x the standing height is obtained. The corresponding values in modern man for 20° and 35° obliquity are 1.27 and 1.46 respectively. A complete set of tracings of the subject were made, and in every case when the limb outlines were sufficiently clear a construction of the axes of the thigh and shank were made. The angles of the segments to the vertical were measured as they appeared on the film. Because of the obliquity of the walk to the image plane of the camera (assumed

> The present analysis suggests that Sasquatch was 6 ft. 5-in (1.98m) in height, with a weight of about 280 lb (127kg.) and a foot length (mean of 4 observations) of about 13.3 in. (34cm.).

to be 27°, the actual angles of the limb segments to the vertical in the sagittal plane were computed by dividing the tangent of the apparent angles by the cosine of 27°. This gave the tangent of the desired angle in each case, from which the actual thigh and shank angles were obtained. The knee angle was obtained as the difference between the thigh and shank angles. A summary of the observations is given in the table shown below.

FRAME NO.	EVENT OR COMMENT	Apparent on film			Corrected for 27° obliquity		
		Thigh	Knee	Shank	Thigh	Knee	Shank
3	R. toe-off	+ 7	14	− 7	+ 8	16	− 3
4		+ 1	19	−18	+ 1	21	−20
5		− 7	10	−17	− 8	11	−19
6	blurred	−18	3	−21	−20	3	−23
7	R. foot pass L.			UNCERTAIN			
8				OF			
9				LIMB			
10				OUTLINES			
11 }	R. heel strike			HERE			
12 }		−27	13	−40	−30	13	−43
13	L. toe-off	−25	22	−47	−28	22	−50
14		0	61	−61	0	64	−64
15		+10	63	−53	+11	67	−56
16	L. foot pass R.	+10	64	−54	+11	68	−57
17		+13	62	−49	+14	66	−52
18		+17	45	−28	+19	50	−31
19		+23	38	−15	+25	41	−16
20		+28	29	− 1	+31	32	− 1
21 }	L. heel strike	+17	6	+11	+19	7	+12
22 }		+20	10	+10	+22	11	+11
23		+19	16	+ 3	+21	18	+ 3
24 }	R. toe-off	+17	18	− 1	+19	20	− 1
25 }		+19	33	−14	+21	36	−15
26		+ 8	15	− 7	+ 9	16	− 7
27		+ 2	19	−17	+ 2	21	−19
28 }	R. foot pass L.	+ 4	28	−24	+ 4	30	−26
29 }				NO MEASUREMENT			

The pattern of movement, notably the 30° of knee flexion following heel strike, the hip extension during support that produces a thigh angle of 30° behind the vertical, the large total thigh excursion of 61° and the considerable (46°) knee flexion following toe-off, are features very similar to those for humans walking at high speed. Under these conditions, humans would have a stride length of 1.2x stature or more, a time of swing of about 0.35 sec., and a speed of swing of about 1.5x stature per second.

Conclusions

The unknown framing speed is crucial to the interpretation of the data. It is likely that the filming was done at either 16, 18 or 24 frames per second and each possibility is considered below.

If 16 fps is assumed, the cycle time and the time of swing are in a typical human combination but much longer in duration than one would expect for the stride and the pattern of limb movement. It is as if a human were executing a high speed pattern in slow motion. It

	16 fps	18 fps	24 fps
Stride length approx.	262 cm.	262 cm.	262 cm.
Stride/Stature	1·27–1·46	1·27–1·46	1·27–1·46
Speed approx.	6·7 km./hr	7·5 km./hr	10·0 km./hr
Speed/Stature	0·9–1·04 sec.[1]	1·02–1·17	1·35–1·56
Time for complete cycle	1·41 sec.	1·25 sec.	0·94 sec.
Time of swing	0·53 sec.	0·47 sec.	0·35 sec.
Total time of support	0·88 sec.	0·78 sec.	0·58 sec.
One period double support	0·17 sec.	0·15 sec.	0·11 sec.

is very unlikely that more massive limbs would account for such a combination of variables. If the framing speed was indeed 16 fps it would be reasonable to conclude that the metabolic cost of locomotion was unnecessarily high per unit distance or that the neuromuscular system was very different to that in humans. With these considerations in mind it seems unlikely that the film was taken at 16 frames per second. Similar conclusions apply to the combination of variables if we assume 18 fps. In both cases, a human would exhibit very little knee flexion following heel strike and little further knee flexion following toe-off at these times of cycle and swing. It is pertinent that subject has similar linear proportions to man and therefore would be unlikely to exhibit a totally different pattern of gait unless the intrinsic properties of the limb muscles or the nervous system were greatly different to that in man. If the film was taken at 24 fps, Sasquatch walked with a gait pattern very similar in most respects to a man walking at high speed. The cycle time is slightly greater than expected and the hip joint appears to be more flexible in extension than one would expect in man. If the framing speed were higher than 24 fps the similarity to man's gait is even more striking. My subjective impressions have oscillated between total acceptance of the Sasquatch on the grounds that the film would be difficult to fake, to one of irrational rejection based on an emotional response to the possibility that the Sasquatch actually exists. This seems worth stating because others have reacted similarly to the film. The possibility of a very clever fake cannot be ruled out on the evidence of the film. A man could have sufficient height and suitable proportions to mimic the longitudinal dimensions of the Sasquatch. The shoulder breadth however would be difficult to achieve without giving an unnatural appearance to the arm swing and shoulder contours. The possibility of fakery is ruled out if the speed of the film was 16 or 18 fps. In these conditions a normal human being could not duplicate the observed pattern, which would suggest that the Sasquatch must possess a very different locomotor system to that of man.

D. W. GRIEVE, M.SC., PH.D.,
Reader in Biomechanics
Royal Free Hospital School of Medicine
London

References

1. Grieve D. W. and Gear R. J. (1966) The relationships between Length of Stride, Step Frequency, Time of Swing and Speed of Walking for Children and Adults. *Ergonomics,* 5, 379-399; Grieve D. W. (1969) The assessment of gait. *Physiotherapy,* 55, 452-460.

2. Green J. (1969) *On the Track of the Sasquatch.* British Columbia Provincial Museum. Cheam Publishing Ltd.

Conclusions Reached by the North American Science Institute (NASI):

Under the direction of J. (Jeff) Glickman, a Certified Forensic Examiner, the North American Science Institute (NASI) performed intensive computer analysis on the Patterson/Gimlin film over a period of three years. At the same time, the institution carried on with general bigfoot research previously performed by The Bigfoot Research Project. In June 1998, Mr. Glickman authored and issued a research report entitled Toward a Resolution of the Bigfoot Phenomenon. The main report findings applicable to the Patterson/Gimlin film are summarized as follows:

J. Glickman

"Despite three years of rigorous examination by the author, the Patterson-Gimlin film cannot be demonstrated to be a forgery at this time."

1. Measurements of the creature: Height: 7-feet, 3.5-inches /2.2m; Waist: 81.3-inches/2.1m; Chest: 83-inches/2.1m; Weight: 1,957-pounds/886.5kg; Length of arms: 43-inches/1.1m; Length of legs: 40-inches/1.02m. (See Note below on height/weight.)

2. The length of the creature's arms is virtually beyond human standards, possibly occurring in one out of 52.5 million people.

3. The length of the creature's legs is unusual by human standards, possibly occurring in one out of 1,000 people.

4. Nothing was found indicating the creature was a man in a costume (i.e., no seam or interfaces).

5. Hand movement indicates flexible hands. This condition implies that the arm would have to support flexion in the hands. An artificial arm with hand movement ability was probably beyond the technology available in 1967.

6. The Russian finding on the similarity between the foot casts and the creature's foot was confirmed.

7. Preliminary findings indicate that the forward motion part of the creature's walking pattern could not be duplicated by a human being.

8. Rippling of the creature's flesh or fat on its right side was observed indicating that a costume is highly improbable.

9. The creature's feet undergo flexion like a real foot. This finding eliminates the possibility of fabricated solid foot apparatus. It also implies that the leg would have to support flexion in the foot. An artificial leg with foot movement ability was probably beyond the technology available in 1967.

10. The appearance and sophistication of the creature's musculature are beyond costumes used in the entertainment industry.

11. Non-uniformity in hair texture, length, and coloration is inconsinent with sophisticated costumes used in the entertainment industry.

Mr. Glickman, shown here with some of his computer equipment, closes his scientific findings with the following statement:

"Despite three years of rigorous examination by the author, the Patterson-Gimlin film cannot be demonstrated to be a forgery at this time."

Personally, I believe Mr. Glickman did an excellent job. The main criticism voiced by many bigfoot researchers was his estimate of the creature's weight. Nevertheless, while we have generally settled on a much lesser figure than his estimate (1,957 pounds/886.5 kg), Mr. Glickman apparently stands firm on his figure as he has not revised it.

Unfortunately, the NASI Report did little or nothing to heighten the credibility of the creature in the eyes of the general scientific community. Full recognition by science demands that there be a body, a part of a body or at least bones. This issue has raised a lot of controversy and has divided researchers on the question as to our right to kill one of the creatures. Up to this point in time, those people who claimed they had the opportunity to shoot a sasquatch did not do so because the creature looked too human.

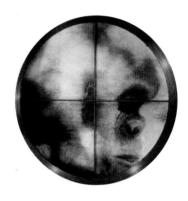

Note on Height and Weight

The actual walking height of the creature in the film has been the subject of considerable controversy over the past five years. Dr. Grover Krantz arrived at a maximum walking height of 72-inches/1.83m; John Green, 80-inches/2.03m; Dmitri Bayanov and Igor Bourtsev, about 78-inches/1.98m; Yvon Leclerc, 75.5-inches/1.92m; J. Glickman (as shown above), 87.5-inches/2.22m, Dr. Donald Grieve, 77-inches /1.96m. In all cases, to determine the height of the creature if it were standing fully erect, we must add something. As the foregoing calculations are based on different film frames, then the specific amount added will differ in all cases. Dr. Krantz estimates that the final figure can be reasonably determined by adding between 8% and 8.5% to the walking height. The weight of the creature at 87.5-inches/2.2m (NASI) is now more con-servatively estimated by Dr. Henner Fahrenbach at 542 pounds/245.5 kg. Whatever the case, the creature filmed was very tall and massive. The illustration shown is by Yvon Leclerc who is seen in the comparison.

FRAME 323 - PATTERSON/GIMLIN FILM Subject Height Calculation

Conclusions Reached by Dr. Grover S. Krantz. Anthropologist, Washington State University:

The following information has been reprinted from the book ***BIGFOOT – SASQUATCH EVIDENCE*** by Dr. Krantz (1999, Hancock House Publishers Ltd.)

Dr. Grover S. Krantz

No matter how the Patterson film is analyzed, its legitimacy has been repeatedly supported. The size and shape of the body cannot be duplicated by a man, its weight and movements correspond with each other and equally rule out a human subject; its anatomical details are just too good. The world's best animators could not match it as of the year 1969, and the supposed faker died rather than make another movie. In spite of all this, and much more, the Scientific Establishment has not accepted the film as evidence of the proposed species. There are several reasons for this reluctance that are worthy of some discussion.

Most of the analyses of the film and its background were made by laymen; their studies and conclusions were published in popular magazines and books, not scientific journals. Most of these investigators did not know how to write a scientific paper or how to get one published. If they had submitted journal articles, these probably would have been rejected simply because the subject was not taken seriously by the editors, no matter how well the articles may have been written. Thus the potentially concerned scientists were simply unaware of the great quantity and quality of evidence. Most of them had heard about the movie, but were reluctant to look into it until someone else verified it. Since they all took this attitude, preferring not to risk making themselves look foolish, nothing much ever happened.

Patterson's was the first movie film ever produced purporting to show a sasquatch in the wild. Since that time many more films have appeared. I have seen eight of them and they are all fakes. A few of the most absurd of these are available on a video cassette. (One other shows a distant, non-moving object that could be a sasquatch, but there is no way to find out for sure.) Given that such faking exists now, it is not surprising that scientific interest in supposed sasquatch movies is even less today that it was back in 1967.

"In many popular publications about the sasquatch there are claimed connections with the truly paranormal, and even fewer scientists want to deal with this."

In many popular publications about the sasquatch there are claimed connections with the truly paranormal, and even fewer scientists want to deal with this. The lunatic fringe has the sasquatch moving through space-time warps, riding in UFOs, making telepathic connections, showing superior intelligence, and the like. All of these enthusiasts try to capitalize on anything new that comes out on the subject. Most of them will eagerly latch on to any scientist who shows an interest, and attempt to lead him/her down their own garden path. It is tantamount to academic suicide to become associated with any of these people.

Finally, and most important, there is the absence of any definitive proof that the sasquatches exist at all. If this had been a known species, the Patterson film would have been accepted without question. But without the clear proof that biologists are willing to accept, a strip of film is of little persuasive value. Of course a film like this would have been accepted as fairly good evidence for a new species of cat or skunk, but even then the type specimen would still have to be collected to make it official. For something so unexpected (at least to science) as the sasquatch, the degree of proof that is required rises proportionally.

What is said here about scientific ignorance regarding the Patterson film is equally true for the footprint evidence and the testimony of eyewitnesses. None of this is normally published in the scientific journals, hoaxes do occur, and the lunatic fringe is all over the place. I don't know of a single scientist who has firmly denied the existence of the sasquatch on the basis of reasonable study of the evidence. Instead of this, most scientists deny it because, to the best of their knowledge, there is no substantial body of evidence that can be taken seriously.

Some of the Russian investigators, not part of their Scientific Establishment, have pushed hard for further study of the Patterson film. Their hope is that such work might establish the existence of these creatures without the necessity of collecting a specimen directly. I wish this were true. Scientific knowledge of the mechanics of bodily motion certainly has advanced in the last twenty years since Donskoy and Grieve studied the film. There are experts in sports, medicine, anatomy, athletics, running shoe design, special effects and prosthetics who could probably make informed judgments on this film. Dmitri Bayanov has urged me and others to pursue these experts, but what efforts have been made along this line have produced no useful results. I can't afford another full round of expert chasing after my episode with the dermal ridges, but at least I have tried. Perhaps someone else will pursue this more diligently in the future. It is not likely that further study of the film can extract any more information than I already have, but it would make an enormous difference if a neutral expert with more appropriate credentials could just confirm what has been presented here.

NOTE: *Mr. J. Glickman, a neutral expert with appropriate credentials did essentially confirm Dr. Krantz's findings as previously presented (NASI Conclusions). The only contentious issues were the creature's height and weight established by Glickman.*

THE MAJOR HOAX CLAIMS REGARDING THE PATTERSON/GIMLIN FILM

At the time of the publication of this work, there have been seven major claims that the Patterson/Gimlin film was a hoax. I discuss each in turn. It might be noted that I do not consider the Ray Wallace claim as a major claim. It is too ridiculous to consider in this category.

1. The John Chambers Connection

The John Chambers connection was a "natural." Chambers had designed the ape heads/faces for the movie Planet of the Apes which was released in 1968, the year following Patterson and Gimlin's experience at Bluff Creek. The apes in the movie were very convincing but totally different from the bigfoot creature seen in the Patterson/Gimlin film. It was, however, reasoned by some people that John Chambers could have, or might have, created the Bluff Creek bigfoot. Apparent rumors made the rounds in Hollywood circles and were picked-up by writers. Why Chambers would have created the creature and who would have compensated him was never speculated. The controversy flared up and died down over the years. Chambers was finally requested to be interviewed on the issue in 1996. At this time, Chambers was in a retirement home. A spokesperson for Mr. Chambers replied to the request informing that Chambers was not available for interviews but went on to state, "He said that he did not design the costume." This statement was taken to possibly mean that someone else designed "the costume" and Chambers made it. Another veteran make-up artist and friend of Chambers added fuel to the fire by stating that Chambers would never admit it if he was involved.

Fortunately, in 1997 the whole issue was laid to rest by Bobbie Short, a registered nurse and bigfoot researcher. Bobbie was granted a personal interview with Chambers and received direct answers to her questions. Not only did Chambers deny any involvement with the Patterson/Gimlin film, but in Chambers' opinion, not he nor anyone else could have fabricated the creature seen in the film. Chambers stated that he was good, but not that good. Chambers admitted that he was aware of rumors concerning his involvement in the film. He never took steps to set the record straight because it was "good for business." One final note -- Chambers had never met or heard of Patterson or Gimlin prior to October 20, 1967. He had never heard of Al De Atley.

2. The Harry Kemball Fiasco

The Harry Kemball fiasco was far less credible than the Chambers connection. In a letter to The 'X' Chronicles dated May

14, 1996, Kemball states he saw Patterson and friends put together a film hoax. Kemball claimed he witnessed this event at the CanWest film facility in North Vancouver, British Columbia. In his own words, Kemball states,

> *I was in the CanWest 16mm Film Editing Room in 1967 when Roger Patterson and friends put together his BigFoot Hoax on 16mm film. They all laughed and joked about the rental of the Gorilla Costume and the construction of the Big Feets. One of his extra tall buddies played the Role of BigFoot. They carefully chose muddy ground so that the foot prints would expand. They carefully shot it on 16mm Kodak EF High Speed Color Positive film stock and when the film is force processed in "Hot Soup" - the film grain is enlarged to add to the sense of mystery. They added a shaky camera zoom with the right amount of "out of focus" to complete the deception. It amazes me that tho frame you published on the front page of your Jan/Feb issue is the same frame that all the Media World-Wide has used over the last 29 years. I sincerely hope that Patterson isn't getting paid for this nonsense! As a graduate in comparative anatomy studies this creature does not and has never existed."*

The 'X' Chronicles admittedly misread the letter and stated in their release that it was one of Kemball's extra tall buddies who played the role of bigfoot. A newspaper tabloid got hold of the story and naturally added more misinformation. The tabloid stated that a person (Kemball) has come forward who claims to have helped create the film. Each of these details, of course, gave unwarranted credibility to Kemball's claims.

Peter Byrne contacted Kemball which resulted in more ridiculous claims by the latter. Kemball incorrectly identified the type of camera Patterson used. He also informed that Patterson had made a death-bed confession to the effect that the film was a hoax. Yet it appears that at the time Kemball wrote the letter, he was unaware that Patterson had died 24 years earlier. When asked why he did not come forward with his claim some 29 years ago, Kimball's reply was surprising. He stated that in the little town of Cranbrook, British Columbia, where he lives, there had been a recent upsurge in crime. There was even a police standoff that apparently shocked the residents of the sleepy town. Kemball reasoned that hoaxes cause crime so he released the information as a crime-fighting gesture.

Moreover, it has been established that the creature in the film is at least 6-feet, 6-inches/1.98m tall. If, as Kemball implies, the creature-actor was with Patterson, surely Kemball would have noticed and given this person more profile than just "extra tall."

Nevertheless, it is reasonable to assume that Kemball based his claim on something. As it happened, René Dahinden (who is five feet six inches/1.7m tall) and John Green (who is well over six feet /1.8m tall), visited the CanWest laboratory in January 1968 to have some work done on the Patterson/Gimlin film (enlargements, etc.). It is very possible staff at the laboratory joked about the film and were overheard by Kemball. Also, it is very likely Kemball mistook Dahinden and Green for Patterson and his "extra tall buddy." There is, however, one minor reference that indicates Patterson was in Canada prior to October 20, 1967. This reference is in *The Times-Standard* article published on October 21, 1967. The articles states: "This year he (Patterson) has been taking films of tracks and other evidence all over the Northwestern United States and Canada for a documentary." The 'X' Chronicles suggested solution to the whole issue was to have Bob Gimlin put into a hypnotic state on a television show whereupon he would be asked questions relative to the film. He would then be subjected to a polygraph test and a psychological stress evaluation. Needless to say, this process did not take place and the Kemball fiasco died a natural death.

3. The Clyde Reinke Disclosure

On December 28, 1998 a television documentary entitled The World's Greatest Hoaxes was aired. The Patterson/Gimlin film was discussed along with other "unexplained" subjects. Clyde Reinke of American Enterprises claims that Roger Patterson was employed by this company to participate in the filming of a fabricated bigfoot sighting. We are led to believe that the resulting film was the famous Patterson/Gimlin footage. Reinke states that as personnel director for American National, he signed Patterson's paycheques. Reinke identified a man, Jerry Romney, who was alleged to have acted as the "creature." Romney was interviewed but he flatly denied any involvement. Nevertheless, all of this information is totally misleading. While Roger Patterson was certainly associated with American National Enterprises, this association did not take place until 1970 - about three years after he obtained his footage of the creature. American National wanted to get their own movie of the creature for a specific production and hired Patterson for this purpose. In 1971, American National abandoned the project and decided to use Patterson's footage for their production. There are possibly two reasons why American National did not choose to use Patterson's movie in the first place. Firstly, they thought they could get their own footage (which appears to indicate they thought the film was genuine). Secondly, they may not have been able to reach an agreement with Patterson for the film rights at that time. The rebuttal information presented here was provided by John Green *who was also retained by American National Enterprises 1970 to assist the company with its project.*

4. The Murphy/Crook "Bell" Issue

This claim involved research by your author that revealed what appeared to be an unusual bell-shaped detail in the creature's mid-section. As the same detail, in my opinion, could be reasonably identified on several film frames, I deemed it to have credibility and reasoned that it could be a possible hoax indicator. A number of sasquatch researchers were informed of the finding in September 1998. One researcher, Cliff Crook, gave the detail full credibility as a hoax indicator and I concurred that this was a possibility. We both searched to identify the detail but nothing was found. Crook then asked to report the find to the media and I informed him that such would be "his call." After the press release (November 1998), analysis of the find was then performed by Dr. Henner Fahrenbach who concluded that the detail was simply "photographic noise." Nevertheless, both Cliff Crook and I continued extensive research to identify the detail. Absolutely nothing was found that even resembles the detail. While I still maintain that the detail exists, I believe it is just debris of some sort caught in the creature's fur. In my opinion, the prominence given this issue by the media was hardly justified.

> "I believe it is just debris of some sort caught in the creature's fur."

5. The Yakima Expose

In late January 1999, some people in Yakima who knew Roger Patterson came forward and denounced the film. They broke long silence as a result of the controversy created by the "bell" issue. The following are quotations from a newspaper article by David Wasson that appeared in the *Yakima Herald-Republic* on Saturday, January 30, 1999.

> Friends and acquaintances of Yakima's famed Sasquatch hunter say the film that for three decades has stood as a paranormal icon to the existence of Bigfoot is actually nothing more than a big joke. "This nonsense has got to stop," said Mac McEntire, a retired salesman from Yakima, who remembers Bigfoot expert Roger Patterson and his partner, Bob Gimlin, laughing about the international uproar their grainy 1967 film created. "You see all these scientists talking about how the creature in that film had to be real, and it just makes me want to puke. It's kind of sad that a lot of people won't believe in God, but they'll believe in something like this."

> In Yakima, the film has long been considered an inside joke for dozens of people who hung out with Patterson and Gimlin in the late 1960s and 1970s. McEntire said he used to throw parties at his house back in his "wilder days," and recalls several of Patterson's close friends and business associated openly laughing at a team of anthropologists that had just declared the creature in the film to be authentic. "I used to just kind of think it was a fun thing, too," he said, "I was thinking, what's wrong with him making a little money? But I started thinking if scientists can swallow something like this, what else are they swallowing." When he read a newspaper article earlier this month

describing the Patterson-Gimlin film as "the gold standard" for Bigfoot sightings, McEntire said he decided it was time to step forward.

Others long have questioned the film as well. Bob Swanson, who now lives near Seattle owned Chinook Press in Yakima back in the mid 1960s and agreed to print 10,000 copies of Patterson's first book, a history of Bigfoot sightings and evidence he believed supported the creature's existence. It was before the famous 1967 film. "I got as excited as the dickens," Swanson recalled. "I fell for it hook, line and sinker." With sluggish book sales and a large printing bill still unpaid, Yakima suddenly became a hotbed of Bigfoot activity, Swanson said with a chuckle. Sightings were being reported throughout the Yakima Valley, and Patterson was never far from the scene.

Swanson said one of his press operators had become friends with Patterson and remembers one morning when he showed up late for work. "I asked him where he'd been," Swanson said, "And he told me, "You haven't heard? There was a Bigfoot spotted out in West Valley."

"Sure enough," Swanson said, "a little while later KIT radio had a deal about it on the news and all my pressmen just started laughing, asking him things like, ' did the suit itch?' He tried keeping a straight face but started laughing too, and finally said something like, "It didn't itch too bad."

Some of Patterson's former acquaintances believe the whole episode was intended to be nothing more that a publicly spoof, but that it spun out of control, eventually taking on a life of its own.

6. The Woodward Client

(In the same article above, the reporter states:)

Now comes what could be the final stomp in the 32-year-old debate. Zilla attorney Harry M. Woodard confirmed Friday he's representing a Yakima man who says he wore the elaborate monkey suit in the Patterson-Gimlin film, and that his client has passed a lie-detector test to prove it.

Woodard described the man only as a 58-year-old lifelong resident of the Yakima Valley who approached him a few months ago after a network news program called questioning authenticity of the 1967 film. The man wanted help negotiating a deal for rights to his story, Woodard said Friday, as well as to explore any legal issues he might face as a result of his involvement in the hoax."

(Note: This statement is confusing. It appears to say that a news program person called the Yakima man. How would this person know to call the Yakima man if his alleged involvement in the film was not known to others?)

I had represented his wife about a year or so earlier on a matter separate for this," Woodard said. "A little while back, when questions about the Bigfoot film started to fester in the media again, he came to see me."

Woodard did not disclose the status of any deals being arranged for his client's story. He did, however, provide a statement from retired

Yakima police officer Jim McCormick, a certified polygraph examiner who administered a lie-detector test Thursday afternoon on Woodard's client. Results of the 75-minute examination showed the man was telling the truth when asked about having worn the bigfoot suit in the 1967 film.

7. The Long Shot

On March 1, 2004, a new book by Greg Long entitled *The Making of Bigfoot, The Inside Story (Prometheus Books, March, 2004)* was introduced on the Jeff Rense radio program. Rense first interviewed two of Long's supporters, Robvert Kiviat and Karl Korff. He then interviewed Long and a Yakima, Washington resident, Robert Heironimus, who claims he was the creature in the Patterson/Gimlin film. Heironimus, was the "Woodard client," previously discussed. We are told that Heironimus was verbally contracted (gentleman's agreement) by Roger Patterson for $1,000 to wear a modified gorilla costume (a costume to which Patterson added breasts) for the film sequence. It is alleged that Bob Gimlin originally contacted Heironimus on behalf of Patterson and fully assisted in producing the film. Practice runs for the film were said to have taken place at Patterson's home, "behind his shed." Heironimus, who states he was never paid by Patterson, claimed he had been living in guilt for 36 years. He never divulged his secret because he was honour bound to Patterson not to do so.

Immediately after the program aired, John Green contacted Bob Gimlin and Mrs. Patricia Patterson who were both distressed and abhorred with the allegations. Gimlin and his lawyer issued a statement to Richard Leiby at the *Washington Post* who was preparing an article on the "story." Leiby states in his article, published March 6, 2004: *Tom Malone, a lawyer in Minneapolis, called us on Friday on behalf of Bob Gimlin, associate of the now dead Bigfoot filmmaker. "I'm authorized to tell you that nobody wore a gorilla suit or monkey suit and that Mr. Gimlin's position is that it's absolutely false and untrue."*

At this writing, the book is being highly criticized by sasquatch researchers.

Bigfoot goes Digital

Many of us have seen the remarkable bigfoot documentaries produced by Doug Hajicek (Whitewolf Entertainment Inc.). The following entry and illustrations provides the inside story on Doug's objective and accomplishments.

High tech equipment and computers are everywhere, even bigfoot creatures have not escaped the electronic trend - they have now gone completely digital. Since no bigfoot body is available for scientists to examine, why not examine a digital one? At least, that was the idea of Doug Hajicek, a filmmaker from Minneapolis, Minnesota. Hajicek was on a standard filming mission in the Northwest Territories, Canada when he and other members of his crew saw enormous man-like footprints going in a straight line over the tundra - they even went over 6-foot (1.8m) tall stunted spruce trees. This experience kindled Hajicek's interest in bigfoot and he has since produced over 7 hours of national televisions shows on the creature. His first production, Sasquatch: Legend Meets Science (which many people say is the gold-standard of all bigfoot documentaries) was created for the Discovery Channel. Following this production, he commenced a 13-week series, Mysterious Encounters, for the Outdoor Life Network (OLN). Hajicek uses both digital technology and forensics to try and solve the ongoing bigfoot mystery. As executive producer for Whitewolf Entertainment Inc., Hajicek has access to many computers and high tech "toys," and also the people needed to operate such equipment.

One of the people is Reuben Steindorf, a forensic animator with Vision Realm. Steindorf was the expert Hajicek chose back in 2001 to work on a long term project to completely animate in full 3D the creature in the Patterson/Gimlin film.

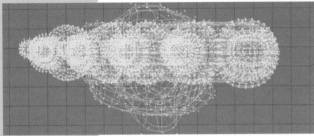

Thousands of data points were used to create the minute details in the toes and feet. This assures as much accuracy as possible.

Reuben Steindorf of Vision Realm works on the feet of the Patterson creature, turning plaster into digital media.

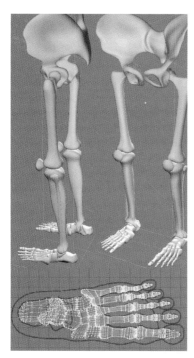
Starting with the feet and lower torso, the animated Patterson-Gimlin creature starts to take shape.

Doug Hajicek (right) and cameraman Mario Benassi prepare to film one of the most high tech forensic reconstructions ever.

Derek Prior, a 3 time All-American sprinter prepares to race a bigfoot near the shores of Lake Chopaka in WA.

Digital motion tests are conducted as the Patterson-Gimlin creature learns to walk. The strange gait is now apparent for the first time in over 37 years.

The digital Patterson creature is literally walking on a virtual exercise wheel so scientists can study the very non-human gait and stride efficiencies.

There appear to be 6 distinct features of the Patterson creature's gait: hip rotation, high leg lift, ankle rotation, non-locking knees, long strides and the legs swing in and out in a criss-cross fashion.

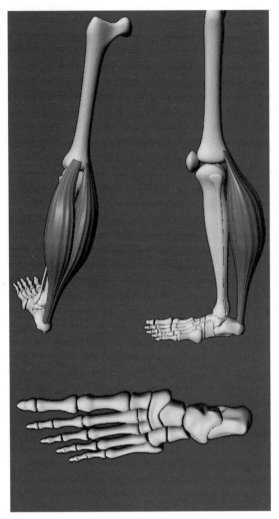

Eventually all of the functioning muscle layers will be affixed onto the bones.

Hajicek knew if Steindorf could effect what he wanted, scientists could study the strange yet graceful criss-cross, hip-rotating, bent-kneed and ankle twisting gait of the creature from any angle they chose. For example, a viewpoint of the Patterson/Gimlin creature walking directly in front of you reveals things that would be otherwise hard to discern in a flat two-dimensional view. With Hajicek's and Steindorf's digital 3D Bluff Creek film site and a 3D creature now complete, any perspective can be custom rendered to fit the study needs of biomechanical experts or scientists.

The 3-year long project has yielded surprisingly accurate results for the study of creature details, right down to its toes. In this connection, Steindorf used actual footprint casts made by Patterson to model extremely accurate feet. Steindorf started with the individual bones and cartilage then worked his way up to muscles, fat, skin - and now even digital hair. Inverse kinematics were used along with motion tracking, locking onto fixed objects in the film, to accurately recreate the creature's movements. Further, Patterson and Gimlin themselves plus their horses were digitally reconstructed.

When the digitized creature was seen walking from another angle, it was obvious that it's legs were operating in a graceful coordinated "swimming type" motion. Hajicek has since coined the term, " The Mountain Gait," to describe this motion.

Moreover, by using forensic technology, a hairless Patterson/Gimlin creature was created that visibly made sense in the physical world to both Steindorf and Hajicek. "It was a bit of a shock to see how human the alleged creature looked without the facial hair," said Hajicek. He also points out that subtle aboriginal features appeared, such as high cheek bones, as the face was carefully reconstructed using a variety of methods.

Digital Film Forensics

The digital Bluff Creek, CA film site can now be viewed from any angle with live action. Steindorf and Hajicek will continue to dial in the animation, adding detail elements.

(Scenes shown here are samples only.)

Hajicek has spent the last 4 years applying technology to the bigfoot mystery in hopes of providing a better understanding of the creature. He knows that his digital work adds only small pieces to a big, unsolved puzzle, "Technology will never replace field studies but it does greatly enhance such studies. Being out in the woods armed with a digital thermal camera is a great feeling, you know nothing can hide from you," he said.

The list of new digital technology processes Hajicek has applied to bigfoot research is extensive, ranging from night vision TV transmissions using a remote controlled helium blimp to digital IR camera traps and most everything inbetween. "This tech stuff works great," Hajicek remarked, "but you still need months in the bush to see the real benefits."

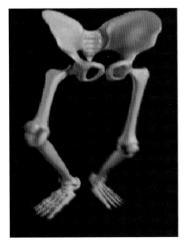  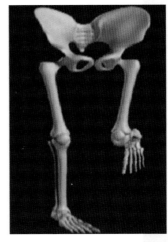

    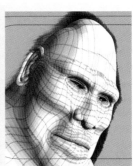

Blow up of frame 352   *Sketch over*   *Full digital face*   *Digital face merge*   *Final digital face*

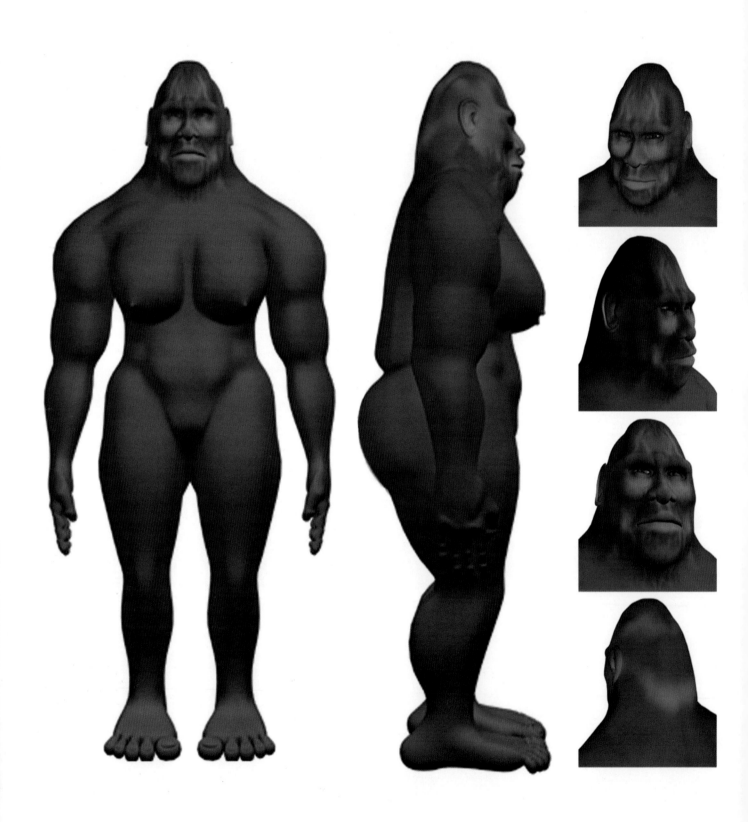

Multiple angle view of a digital bigfoot without hair.

Digital Equipment for Bigfoot Research

CSI forensic techniques were applied during the reconstruction of the "Memorial Day" footage shot by Lori and Owen Pate in 1996. The challenge was to determine how fast the creature was running, calculate its size and determine if a human being could duplicate its speed. The process involved using fixed objects in the film site background to establish the speed of the creature. Hajicek hired 3-time All-American sprinter, Derek Prior, to try and beat the big ape in a race. Before the race could begin, the exact path the creature took needed to be staked out. The next step was to have Pacific Survey & Supply scan the entire mountain with a Cyrax digital lidar radar scanner. Prior was then fitted with a GPS and laser tracker to measure his speed and any path deviations. Even vertical movement was recorded as he ran - then back to the lab to analyse the results. We won't give away the findings here, they can be found in *Sasquatch: Legend Meets Science* (DVD or the book).

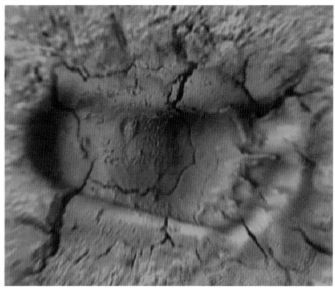

Tracks can now be digitally captured in 3D using hand-held scanners, replacing plaster altogether.

Digital map of bigfoot sightings.

A virtual 3D plaster cast can be studied in great detail.

This handheld scanner could replace plaster in the future, even allowing accurate virtual casting of snow tracks.

The path the Memorial Day creature took is noted from this digital survey.

Lidar laser scanner

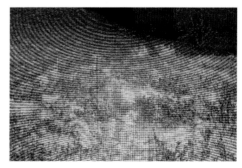

300 million data points were gathered just because of a bigfoot sighting.

FOR THE RECORD - SASQUATCH INSIGHTS

I present here some remarkable sasquatch artwork and related artifacts together with some insights on the Patterson/Gimlin film and the sasquatch in general.

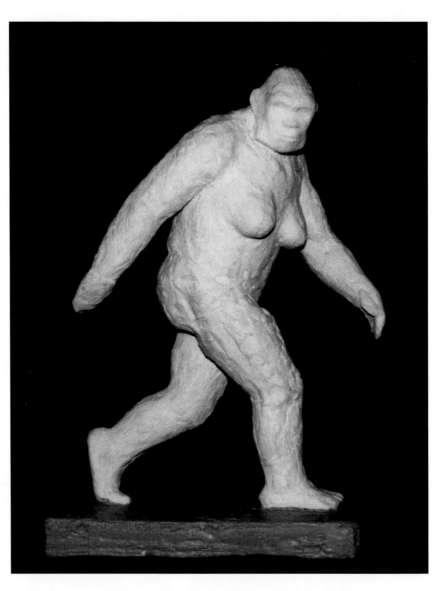

This sculpture (to scale) created by the Russian hominologist, Igor Bourtsev, shows the creature in the Patterson/Gimlin film as it turns and looks at the two men. Igor is shown at left with his sculpture. Also shown is the film frame image (frame 352) used as the basis for the sculpture.

The following is a painting of a sasquatch by Robert Bateman, the noted wildlife artist. I do not know on what the image is based, however, the creature's stance appears to have similarities to frame 352 in the Patterson/Gimlin film (also, in subsequent frames the creature does go behind a tree). Bateman has certainly provided an intriguing and beautiful "atmosphere" for the creature. He gives us the feeling of a great "mystery," which is highly appropriate.

Painting by Robert Bateman

My main concern with the painting is the creature's gorilla-like nose. It does not appear to correlate with what we see (or think we see) in the Patterson/Gimlin film, resulting in a more ape-like than human-like appearance. Nevertheless, many prominent researchers are of the opinion that the creature is, in fact, just another member of the great ape family. If so, then perhaps Bateman's painting is reasonably accurate. I must admit that the film is really not that clear on the creature's facial features.

The following are artistic conceptions created by Yvon Leclerc. Yvon has used the Patterson/Gimlin film for his insights.

In this image we perceive a far more human sasquatch. It is interesting to note that one of the West Coast First Nations' names for the creature is "Gilyuk" which means, "the big man with the little hat." This name was derived because from a distance the creature's pointed head gives the appearance that it is wearing a little hat. Yvon created this image using the upper skull of a gorilla and the lower skull of homo erectus (Peking man). Note the sagittal crest.

There is no evidence that the creature uses fire and it is not believed it uses any sort of coverings. Like all wild creatures, therefore, it would probably have very thick hair for warmth and might appear somewhat menacing as seen here.

There have been a number of sightings that involve sasquatch babies and children. Certainly sightings of this nature serve to further attest to the creature's existence. The fact that the creature in the Patterson/Gimlin film was a female sparked the creation of this little scene. We can certainly speculate that somewhere in our vast wilderness there are mothers tending to their children just as we see here.

The creature in the Patterson/Gimlin film appears to have some distinct head and facial characteristics as seen in the following illustrations.

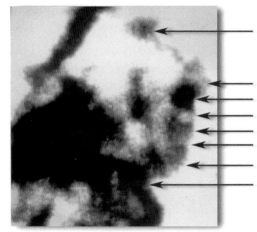

Very high, sloped head (Possibly a sagittal crest)

Heavy brow ridge
Virtually no nose bridge
Small, but wide "PUG" nose
Wide area - nose/upper lip
Thin or "normal" lips
Virtually no chin
Virtually no neck

frame 339 - Patterson/Gimlin film

This sculpture might provide some additional insights as to the appearance of male sasquatch. I am inclined to think that males would have more facial hair than females and that during the winter, both males and females would have thicker hair. We must consider that this creature does not use fire and therefore hair would provide its main means to keep warm. Northern Canada and Alaska, for certain, get very cold.

One might note that the finished sculpture shown does not appear to have a highly pronounced sagittal crest or "pointed head." The extent

of this feature apparently has a lot to do with hair and camera angle. This photograph shows the sculpture before hair was added and other detailing performed. Even here, however, the "point" is not as severe as we see in frame 339, previously illustrated. I believe "point" severity depends more on hair than it does head (a conclusion reached by Dmitri Bayanov some years ago). Whatever the case, is does appear that sasquatch have an unusually large sloped head.

The size of the creature's head appears to fit about five times into its walking height. We need to add one more head (maximum) to account for additional height if the creature were standing fully erect (standing height). We therefore appear to have a ratio of 6 to 1. With adult human beings, the same ratio is 8 to 1. While this analysis is very rough, it does indicate that sasquatch might have a much larger head in relation to body height that that of humans.

The following is an eye-witness drawing of a sasquatch (left) and an enhancement of same by Yvon Leclerc. The witness claims many sightings of the creatures, both adults and children, over a number of years. The drawing is of a very young sasquatch. Remarkably, here we have the same sort of nose as that envisioned by Robert Bateman.

*Current technology enables artists and engineers to make extremely life-like props for the motion picture industry. Wild Entertainment Inc., and Wilderness Productions Inc., the producers of the movie **Sasquatch** released in 2003, had a sasquatch arm fabricated, as seen here, with moveable fingers controlled by wires inside the arm (cost: $10,000). While impressive, I think the hand is too "human" from what we know.*

In October 1990, Canada Post issued the postage stamps seen here (Canada's Legendary Creatures) in its Canadian Folklore series. The Patterson/Gimlin film was instrumental in the inclusion of the sasquatch stamp (enlarged below). The literature on this stamp references the film.

NOTE: Postage stamps shown are copyright Canada Post Corporation, 1990. Reproduced with permission.

In 1995, bigfoot made the cover story of the Scott Stamp Monthly. I was working with René Dahinden at the time and had him autograph a number of sasquatch postage stamps. The stamp shown here in the upper left corner of the cover is an autographed stamp. The story in the magazine was written by my son, Daniel.

THE PHYSICAL EVIDENCE AND ITS ANALYSIS

While we have yet to capture a living sasquatch or find a body or bones, many people are of the opinion that the evidence we have is sufficient for full recognition of the species. Alternately, it is felt that the evidence is sufficient to at least justify a full government-sponsored inquiry into the issue.

8

FOOTPRINTS AND CASTS

Together with sightings and the Patterson/Gimlin film, footprints and other possible physical evidence serve to indicate that the creature actually exists. Further, this evidence provides insights into the actual size of the creature – both its height and other body measurements.

Large human-like footprints have been found in remote areas all across North America. They are often deeply imbedded in the soil indicating that the creature that made the print was extremely heavy.

Numerous plaster casts have been made of probable sasquatch footprints. They have been studied by many professionals and deemed to be authentic. In other words, they were made from prints created by a natural foot. The following museum exhibit presentation provides an explanation on footprint casting.

Footprint Casts

(The following is the text associated with this photograph.)

How footprint casts are made and what they represent: *Footprint casts are made by pouring plaster directly into a footprint. The plaster flows into all depressions without disturbing even the most minute foot crevices created in the soil or sand. In some cases, dermal ridges (like finger prints) have been found on footprint casts. The person seen here is Roger Patterson making a cast of a footprint left by the sasquatch he filmed at Bluff Creek, California in 1967.*

Plaster takes about twenty minutes to solidify. When the cast is removed from the print, the result is a plaster representation of the UNDERSIDE of the foot. In other words, it is a view of the foot from beneath, not above, as illustrated in the adjacent photograph.

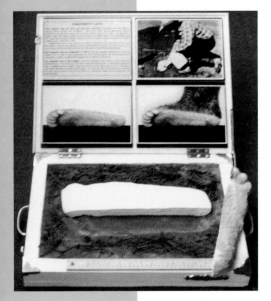

How footprint casts are duplicated: *Footprint casts are generally duplicated by using the original cast to make a sand impression and then pouring plaster into the resulting print. Alternately, a mold is made of the original cast for plaster reproductions.*

The footprint casts in this exhibit: *Most of the footprint casts shown in this exhibit are duplicated casts. They were produced from either the original cast or a subsequent generation copy. Original casts are slightly larger than the actual foot that made the print. Duplicated casts made with sand are slightly larger again. Some of the casts in this exhibit are estimated to be up to 1.1-in./2.8cm larger than the actual foot that made the print.*

NOTE: What is stated here also applies to the casts shown in this book.

"Original casts are slightly larger than the actual foot that made the print. Duplicated casts made with sand are slightly larger again."

What might a Sasquatch Foot Actually Look Like?

The following photographs show a sculptured clay foot that is based on a 1958 Bluff Creek, California cast (print found by Bob Titmus). An actual plaster cast forms the sole of the foot. The human foot shown for comparison is 11.5-in./29-2cm long.

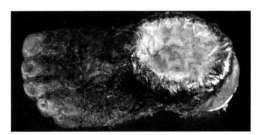

Cast Considerations: When a foot is pressed into a soft surface such as soil or sand in the act of moving, three processes take placed that affect the size of the impression made by the foot. First, the movement or motion of the foot cause some "slide" or "skid." Second, the foot itself expands slightly in all directions. (It might be noted that for this reason one always "tries out" a new pair of shoes - weight placed on feet causes them to spread out). Third, the foot marginally displaces the sand or soil. It is impossible for an impression to be exactly the same size as the object that made

Cast-Making Box

This is a cast-making box. It has hinged lockable lids on both the top and bottom. One lid is shut and locked. The cast to be duplicated is placed "face up" in the box. Sand is then placed (gently pressed) on top of the cast, filling the box to the absolute brim. The open lid is then shut and locked and the box is turned upside down. The other lid is now opened, revealing the cast fully immersed in the sand. The cast is then gently removed leaving a perfect impression for casting (i.e., pouring of plaster into the impression).

the impression. One can prove this by trying to fit two circular objects of exactly the same diameter into one or the other.

On the right the first photograph (left position) is a perfect cast of my own foot. I made the imprint from which the cast was made by pressing my foot into sand. To get my foot firmly into the sand deep enough for a cast imprint, I had to use some motion (press down with my weight a few times and "jiggle" a little). This motion would partially be the same as walking motion, but not nearly as severe. After making the cast, I trimmed it to the exact (as close as possible) outline of my foot. I then made a transparency of my  foot with no weight on it using a photocopier (i.e., I placed my foot directly on the photocopier plate and took a color photocopy.) The second photograph (right position) shows the transparency positioned on the back of the cast. The white margin around my foot is the amount of cast expansion cause by the conditions mentioned. It should be noted that not only is the cast longer and wider than my foot but all details within the cast are larger (compare the relative size of the toes). It appears my second toe (from left) pushed out more than the others, causing a wider discrepancy.

It has been reasoned that the foot of a sasquatch would have a very thick pad. The illustration seen here of a possible sasquatch foot offered by Dr. Jeffrey Meldrum provides some insights. I believe a foot of this nature would spread out considerable with the excessive weight of these creatures - much 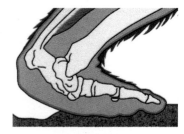 more than a bony human foot. For that reason **alone** I believe **original** footprint casts are larger by up to .5-in./1.27cm in all directions. When we add slide and soil displacement, we need to add up to another .20-in./5mm. We are therefore up to .70-inches/1.8cm difference between the actual footprint and the actual foot with no weight on it. A cast made from the print will naturally be up to this amount larger.

When casts are duplicated by pressing them into sand, only movement and soil displacement affect size because the cast is solid. A first generation cast would probably increase by up to .12-inches/3mm. When casts are serially reproduced, this additional enlargement factor is compounded.

Casts made from moulds, of course, do not "grow." Further, casts made with a cast-making box whereby the cast is not moved or pressed in the recasting process have insignificant growth.

FOOTPRINT CAST GALLERY

The following gallery of footprint casts provides some insights as to the different foot sizes and shapes. Sasquatch, it appears, are just as varied as human beings in their physical makeup.

NOTE: Refer to the previous section for information on cast growth. The adjacent chart provides statistics.

CAST GROWTH COMPARISON TO ACTUAL FOOT - NO WEIGHT	
CAST GENERATION	LARGER BY (MAX.)
ORIGINAL CAST	.70-inches (1.8cm)
1st GENERATION	.82-inches (2.1cm)
2nd GENERATION	.94-inches (2.4cm)
3rd GENERATION	1.1-inches (2.7cm)
NOTE: The increase applies to both the length and the width of the cast and all details within the cast are increased proportionately. Cast generation growth applies only to casts made by *pressing* the cast to be duplicated into sand.	

1. Bluff Creek, California, Jerry Crew, 1958 (2nd generation cast, 17.5-inches/44.5cm long). This is a copy of the famous cast Jerry Crew took to a newspaper and the resulting article gave birth to the word "bigfoot" as the name of the creature in the United States.

4. Believed to be from Bluff Creek, California. The person who made the cast is not known. It was, probably made in the late 1960s, (possible original cast, or 1st generation, 14.5-inches/36.8cm long).

2 Blue Creek Mountain road (Bluff Creek area, California), John Green, 1967 (original cast, 15-inches/38.1cm long).

5 Strathcona Provincial Park, Vancouver Island, British Columbia, Dr. John Bindernagel, 1988 (1st generation casts, 15-inches/38.1cm long). The horizontal lines on this cast were caused by a hiker who stepped in the foot print.

3. Blue Creek Mountain Road (Bluff Creek area, California), John Green, 1967 (original cast, 13-inches/33cm long).

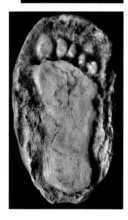

6. Abbott Hill, South Olympic Peninsula, Washington, A.D. Heryford, 1982 (2nd generation cast, 15-inches/38.1cm long). Certainly one of the best casts ever obtained. The copy seen here was professionally produced from a mold by Richard Noll, Edmonds, Washington.

7. Shawnee State Park, Ohio, Joedy Cook, June 18, 2003 (original cast, 15-inches/38.1cm long). A man and his wife found the prints and called a bigfoot hotline. Cook responded and found nine footprints.

8. Chilliwack River, British Columbia, Thomas Steenburg, 1986 (2nd generation cast, 18.5-inches long/47cm). Steenburg was informed of a sighting in the area three days after the occurrence and went to investigate. He independently found 110 footprints 18-inches/45.7cm long.

9. Bluff Creek, California area, Laird Meadow Road, Roger Patterson, 1964 (3rd generation cast, 16-inches/40.6cm long). Prints were found by Pat Grave, October 21, 1964 who told Roger Patterson of the location. The creature that made the prints is believed to be the same as the one that made the prints found by Jerry Crew (see No. 1).

10. Bluff Creek, California. Bob Titmus, 1958 (2nd generation casts, 16-inches/40.6cm long). Both casts are from the same trackway.

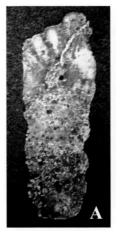
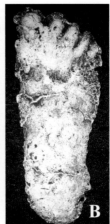

11. Hyampon, California, Bob Titmas, 1963. Hyampom is a tiny village about 60 miles/96.5km south of Bluff Creek. All prints from which these casts were made were found on the same occasion but only the first three prints (casts A-C, which were from the same trackway), were found in the same place. The other two casts (D and E) were from prints found in and additional two separate locations.
A. Original cast, 16-inches/40.64cm long
B. Original cast, 17-inches/43.2cm long
C. Original cast, 16-inches/40.6cm long
D. Original cast, 16-inches/40.6cm long
E. Original cast, 15-inches/38.1cm long

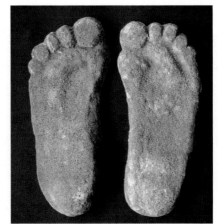

12. Skeena River Slough, Terrace, British Columbia, Bob Titmus, 1976 (2nd generation casts, 16-inches/40.6cm long). Both casts are from the same trackway. Children found and reported the footprints; Titmus investigated and made the casts

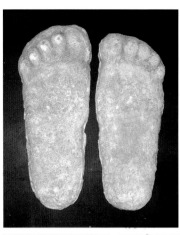

13. Bluff Creek, California (film site), Roger Patterson, October 20, 1967 (2nd generation casts, left cast 15-inches/38.1cm; right cast 14.6-inches/37.1cm long). Actual footprints in the soil measured 14.5-inches/36.8cm.

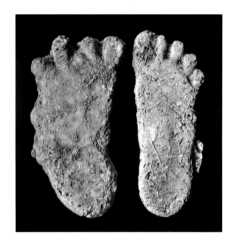

15. Bossburg, Washington, "cripple foot casts," René Dahinden, 1969 (original casts, left cast, 16.75-inches/42.6cm; right cast 17.25-inches/44.5cm long). Over 1,000 footprints were found.

14A-E (See below.)

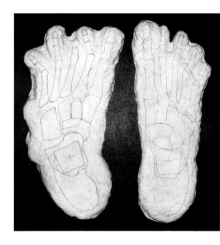

16. This set of the Bossburg "cripple foot" casts shows the speculated bone structure of the feet as determined by Dr. Grover Krantz. It is reasoned that the deformed foot (left cast, but actual right foot of the creature) was the result of an accident or a birth defect. The footprints from which the casts were made were discovered on two different occasions. On the first occasion, a few prints were found and then a few weeks later a long line of prints. The casts were intently studied by Dr. Krantz who was adamant that they appear to have been made by a natural creature. He reasoned that if the footprints were a hoax, then the hoaxer had to have an in-depth knowledge of anatomy. Further, this person would have had a remarkable skill in constructing or designing some kind of apparatus to make the footprints. (See special discussion below.)

14A-E. All of the casts seen here are from the Bluff Creek, California Patterson/Gimlin film site. They were made by Bob Titmus from prints that were still in place on October 29, 1967, nine days after the filming. All casts are originals. They vary in size (14 to 15-inches/35.6 to 38.1cm) in accordance with foot placement and motion.

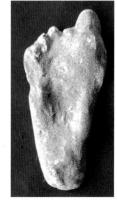

17. Elk Wallow, Walla Walla, Washington, Paul Freeman, 1982 (3rd generation cast, 14-inches/35.6cm long). The cast has an indent about center caused by a rock which the creature stepped on. This cast is a copy of one of three casts made by Paul Freeman on which Dr. Grover Krantz discovered dermal ridges (like finger prints). (See special discussion page 109).

The Jerry Crew cast (No. 1 above) was cleaned-up (sanded/smoothed out) by Bob Titmus and donated to the Vancouver Museum. On the back of the cast there is the following notation written by Titmus. **"This is an actual cast of Bigfoot imprint made Oct. 2, 1958 in Bluff Creek in Humboldt County, California. 'Bigfoot' is not a hoax."** *Bob Titmus, Taxidermist, Anderson, Calif.*

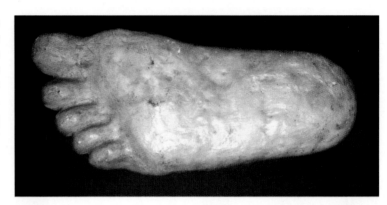

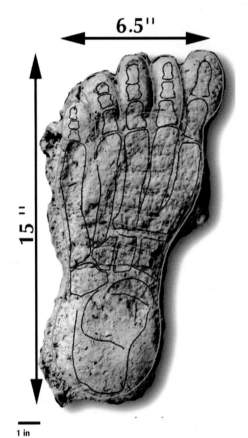

Bone structure for the Strathcona Provincial Park print (Cast #5) as determined by Yvon Leclerc.

Feet illustrations by Yvon Leclerc.

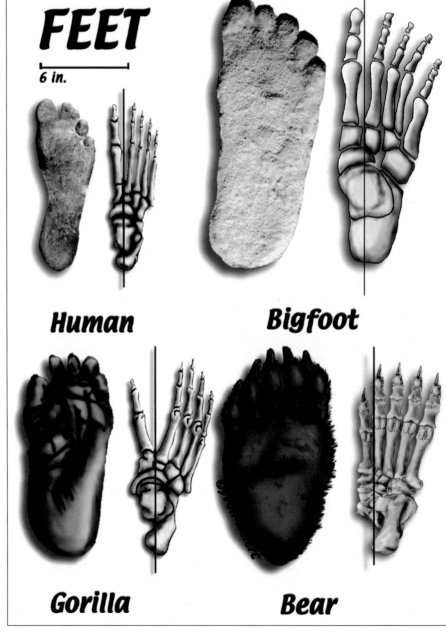

Special Discussion on Cast Credibility: In determining the authenticity of any evidence, be it footprints, handprints or anything else physical in nature, scientists and professional people have only one criterion of examination - *"What does the evidence itself indicate."* They are not swayed in their decision by *"Circumstantial evidence."* If, for example, the physical or hard evidence establishes a hard fact or reasonably hard fact, providing testimony (even sworn testimony) to the contrary is not admissible. Testimony, hearsay, confessions, even photographs are not hard evidence. Really, we would not want this situation to be any other way. Imagine what would happen if judges and juries in our courts gave what people say equal weight with the hard evidence presented in a court case.

In the field of sasquatch studies, we have come face to face with a specific irreconcilable situation: Many researchers do not give any credibility to findings by both Ivan Marx and Paul Freeman. In other words, the researchers consider such findings hoaxed or fabricated. They base their conclusions on either personal contact with Marx and Freeman or information concerning them. Without doubt, Ivan Marx was a notorious practical joker and Freeman's "luck" in finding sasquatch footprints and handprints goes far beyond reason. **However, some of the material they have presented has reasonably or totally withstood all scientific scrutiny.** It is bordering on the impossible to determine how they would manage to hoax or fabricate such material.

In the case of Ivan Marx, the Bossburg "cripple foot" prints were just too "good" and perhaps too numerous for him, or anyone else for that matter, to have fabricated. In the case of Paul Freeman, some of his casts show dermal ridges (like finger prints) that are not human in nature - they appear to be those of a non-human primate. To conclude that Freeman fabricated such dermal ridges goes beyond reason. Nevertheless, the fact remains that there is a possibility (not probability) that both Marx and Freeman found some way to fool the scientists (they would certainly not be the first to do so).

In compiling this work, I have had to come to grips with this issue. John Green and Tom Steenburg were against the inclusion of material found by both Marx and Freeman. They have reluctantly agreed to such inclusions on the basis that I cannot overrule the findings of scientists and professional people. **Let the record show the Green and Steenburg objections.** CLM

"However, some of the material they have presented has reasonably or totally withstood all scientific scrutiny."

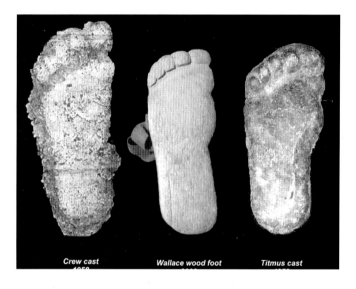

Footprint casts of the quality I have presented are nearly impossible, if not impossible, to produce from prints made with a "fabricated" foot. This illustration prepared by Richard Noll compares sasquatch casts with a wooden foot owned by the infamous hoaxer, Ray Wallace. One can see that there are few similarities. To even consider that Wallace, or anyone else for that matter, could produce convincing sasquatch footprints with a piece of carved wood is absurd.

FOOTPRINT AND CAST ALBUM

Numerous photographs of sasquatch footprints and resulting casts have taken by sasquatch researchers and other people over the past fifty years or so. The following is a presentation of a reasonable cross-section of such photographs. I have also included here some feet comparisons between sasquatch and bears.

Note: The first seven photographs presented were taken of prints found beside the Blue Creek Mountain road, California, in August 1967.

This photograph taken by René Dahinden is considered one of the best ever taken of a sasquatch footprint. The 13-inch/33cm print was in deep dust, dampened on the surface by a brief rain. (Blue Creek Mountain, 1967)

This print is highly similar to the previous print, however, it is two inches longer (15-inches/38.1cm) and was found some distance away. (Blue Creek Mountain, 1967)

The prints seen here are of a 15-inch/38.1cm and a 13-inch/33cm print crossing each others path. (each print is one in a series). In all, 590 prints were counted. However, prints on the traveled part of the road had been obliterated so it is estimated that the actual number was probably well over 1,000. (Blue Creek Mountain, 1967)

Don Abbott of the British Columbia Provincial Museum is seen here attempting to lift a glue-treated print out of the ground. Unfortunately, Don was unable to remove the print intact so it never made it back to British Columbia. (Blue Creek Mountain, 1967)

John Green is seen here measuring the creture's toe to heel pace. Green is using a yard stick and we can see that the prints are about one yard, or 3-feet/91.4cm apart. A 6-feet (1.83m) man would have an equal toe heel pace of 20 to 22-inches (50.8-55.9cm). (Blue Creek Mountain 1967)

This print measures 15-inches/38.1cm long, 7-inches/17.8cm across the ball of the foot and almost 5-inches/12.7cm across the heel. An identical print was first observed and cast nine years earlier. (Blue Creek Mountain, 1967)

Footprint with a boot print, about size 12 boot. (Blue Creek Mountain, 1967)

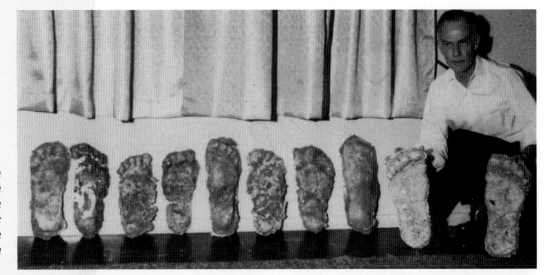

Bob Titmus is seen here with a selection of casts he made from sasquatch prints found in Northern California from 1958 to 1967.

These 17-inch/43.2cm prints were discovered north of Ellensberg, Washington on November 6, 1970.

A man with a size 14 boot compares his foot with a 17-inch/43.2cm cast of a sasquatch print.

Bob Titmus and Syl McCoy with 17, 16, and 13-inch/43.2, 40.6 and 33cm casts.

Sasquatch prints (center line) and human prints on the sand of the Nooksack estuary, Washington, 1967. The following is John Green's account:

"The Nooksack River gets its start in life on the slopes of the highest mountains in northwest Washington, but it runs about 20 miles/32.2km through flat farmlands before it gets to the sea. There is an area of heavy forest on the Lummi peninsula, although it is cut up with roads and there are many houses. There is also heavy growth, and no roads or houses, on the islands in the mouth of the river. It isn't an area that could be expected to house a population of sasquatches on a permanent basis, but if they used the river for a highway, as the Indians say they do, they could easily come down at night and settle in for the fishing season. Most of the 1967 sightings took place in September, and more than half of them were by fishermen drifting with gillnets down the channels at the mouth of the Nooksack. Mr. and Mrs. Joe Brudevoid told me that they had seen an eight-foot/2.44m, black animal with a flat face standing in the river in the early afternoon. It was about 200 yards/182.9m away, and although the water was only up to its knees it bend down and disappeared in it. The river is muddy, so that neither salmon nor sasquatch could be seen beneath the surface, but I was told that sometimes a surge would travel along the river as if something very big was swimming by. In the area of the Brudevoid sighting tracks were later found coming out of the river onto a sandbar and covering about 150 yards/137.2m before re-entering the water. They were 13.5 inches/34.3cm long and sank in two inches/5.1cm. They were flat, had five toes, and took a 45-inch/1.14m pace.

Some casts show evidence of dermal ridges (skin prints). Dr. Grover Krantz discovered this evidence and thoroughly researched his findings with fingerprint experts. In this highly magnified section of a footprint cast, ridges are very clear and close examination reveals tiny holes in the ridges. These holes are believed to be active sweat pores. Dr. Krantz provided casts for examination to over forty experts throughout the world, including experts at the Smithsonian Institution, U.S. Federal Bureau of Investigation (FBI) and Scotland Yard. Opinions ranged from "very interesting," "they sure look real" to "there is no doubt they are real." The only exception was the F.B.I. expert who said, "The implications of this are just too much; I can't believe it (the sasquatch) is real."

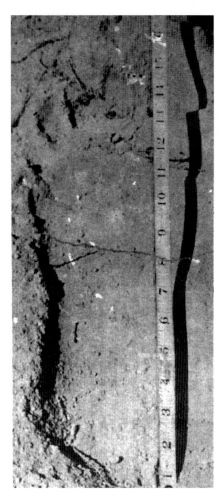

John Green is seen here in 1972 with his collection of footprint casts. John was, and continues to be, the pre-eminent sasquatch investigator and chronicler.

Bob Titmus was told that this is possibly the oldest photograph of a sasquatch footprint. It was taken in 1947 on a utility right of way between Eureka and Cottonwood, California.

The Abbott Hill footprints, one of which resulted in this remarkable cast, (Cast Gallery, Cast #6,) were investigated and cast by Deputy Sheriff Dennis Heryford on April 22, 1982. Abbott Hill is a large tract of land in a fairly secluded area of the eastern portion of Grays Harbor County, Washington. The print used for this cast was 15.5-inches/39.4cm long. Heryford also investigated that same day more prints found at Workman's Bar which is about seven miles from Abbott Hill. These prints, which were of two different lengths, 17 and 15.5 inches/43.2cm and 38.1cm, started from under water. Five days later, more tracks were reported and investigated at Elma Gate which is about 9 miles/14.5 km from Abbott Hill. These prints were 15-inches/38.1cm long. On May 23, 1982, more prints were found at Porter Creek which is in the same vicinity – fewer than 9 miles/14.5 km from Abbott Hill. All of this information is from the official police report on the incidents (no size is shown for the Porter Creek prints).

Two 15-inch/38.1cm prints found on a Bluff Creek, California gravel bar in about 1960. The prints have been sprinkled with white powder for contrast. A pair of tin snips, 10.5-inches/26.7cm long, was placed near a print to provide a sense of the print's length.

A 13-inch/33cm print found on a sandbar beside Bluff Creek in 1967. This photograph gives us a good appreciation of the depth of footprints.

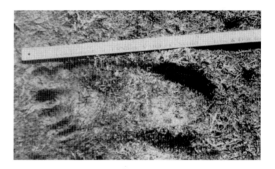

All of the casts seen here, being demonstrated by Bob Titmus, were from footprints found in April 1963 at two sites near a tiny village called Hyampom - about 60-miles/96.5km south of Bluff Creek, California. The casts all measure around 16-inches/40.6cm in length.

Bob Titmus is seen here measuring the Jerry Crew cast (Bluff Creek, 1958; Cast #1 in the Gallery). When Crew decided to make a cast, he contacted Titmus who gave him directions on cast-making.

Actual Hyampom footprints. The first photograph shows a print in wet ground.

Bruce Berryman, Bob Titmus and Syl McCoy display casts of footprints found in Hyampom, California in 1963.

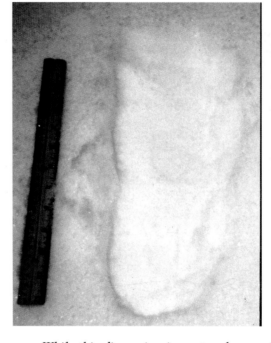 

While this discussion is centered on major findings on the West coast, numerous other footprints have been found and cast throughout the rest of North America. The story associated with this photograph (left) of a 14.5-inch/36.8cm footprint in snow found in Ohio, is very amusing. The photograph was taken by a lady schoolteacher who saw a sasquatch cross the road ahead of her while driving near Hubbard, Trumbull County, in January 1997. Unfortunately, she was not quick enough with her camera to get a shot of the creature. Nevertheless, she courageously stopped her car, got her ruler (which she would naturally have with her), stepped out and took the photograph. If we compare this print to the Titmus 1968 Bluff Creek Cast (left foot) it is seen that they are remarkably similar in shape.

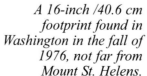

John Green is seen here holding a "cleaned-up" copy of the Jerry Crew cast (Bluff Creek, California, 1958). The footprints found by Crew were quite highly defined because of the soft soil and the creature's great weight. Crew's cast and subsequent copies were therefore also well defined. Copies may be sanded or "detailed" to produce a closer resemblance of the sole. Nevertheless, as a general rule, researchers do not detail casts other than general "clean-up."

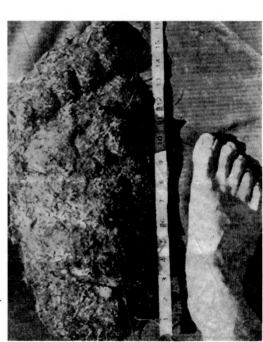

A 16-inch /40.6 cm footprint found in Washington in the fall of 1976, not far from Mount St. Helens.

Dr. Grover Krantz examines one of the Bossburg Washington "cripple foot" prints in snow, December 1969. Dr. Krantz was highly impressed with the casts made from the prints. He stated that the nature of the creature's deformed foot was such that if the prints were a fabrication, then whoever made the prints had to have a superior knowledge of anatomy. Such knowledge, he claimed, was far beyond that of non-professional people. I will mention here that Ivan Marx, the person who first found the prints, was not known to have had knowledge of this nature. Nevertheless, he could have known someone with a deformed or distorted foot and patterned a fabricated foot accordingly for making prints. Moreover it is possible Marx conspired with another person with professional knowledge, or another person with such knowledge fabricated the prints and Marx simply found them as he stated. Opinions remain highly divided on the authenticity of the "cripple foot" prints. In my own opinion if the prints were fabricated, the idea to make one foot deformed was marvelous - perhaps a little too marvelous?

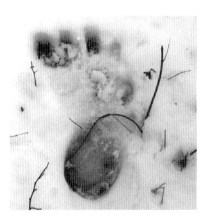

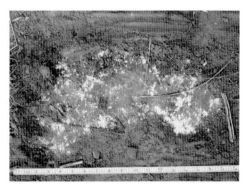

Photograph of a "cripple foot" print in soil. It appears prints in this series were used to make the second cast set

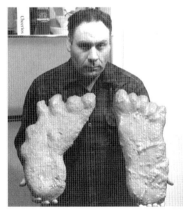

Seen here are the first set (left) and second set (right) of casts made from the unusual Bossburg "cripple foot" prints. It is immediately seen that the deformed foot is far more twisted in the second cast set. Also, the little toe is much straighter. These conditions might indicate that the foot that made the prints had to be very flexible. I have mentioned opinions are divided on the authenticity of the "cripple foot" casts, however, the variation seen here makes fabrication of the prints somewhat harder to explain. René Dahinden is on the left and John Susemiehl, a border patrolman, is on the right.

Norm Davis (left) and his wife, Carol (owners of a Colville, Washington radio station) together with Joe Rhodes inspect "cripple foot" prints near a Bossburg garbage dump. This group made the first casts of footprints found on this occasion.

This footprint, measuring close to 17-inches/43.2cm long and 7-inches/17.8cm wide, was found in March 1960 on Offield Mountain, which is near Orleans, California. (Pacific Northwest Expedition finding.)

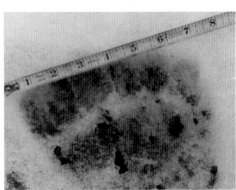

One of several 15.5-inch/39.4cm footprints in a series found in July 1976 along a Skeena River Slough (near Terrace, British Columbia). Young boys found the prints; Bob Titmus investigated and made casts of both the left and right feet. The pair of casts he produced (shown in the Footprint Cast Gallery #12) is a superb example of matching sasquatch footprint casts.

Bob Titmus provided this write-up and photograph relative to the Skeena River slough footprints. Bob was a very methodical and exacting person. He was one of the most highly regarded researchers in the field of sasquatch studies.

> Sasquatch tracks crossed over this pile of stumps & root systems near slough just off Skeena River, near the Terrace, B.C. area. Tracks were 3 or 4 days old & had been exposed to heavy rain a couple of days before being cast & photographed on the evening of July 17, 1976. Tracks measured 15½" long, 6½" wide at the ball & 4" wide at the heel. Walking stride from toe to heel was 78". Heel depth approx. 1⅞" - Toe depth approx. 1⅛". See other photos & casts. 5 casts made in all of the 12 or 15 tracks. Bob Titmus

Bob Titmus holding his freshly-made Skeena River casts. On the right is a photograph detail enlargement of the cast he is holding in his left hand (right facing). To me it is "far-reaching" to conclude that the original footprints were made by anything other than natural feet.

Footprint found in August 1967 on Onion Mountain which is west of Bluff Creek, California. The print measured between 11 and 12 inches/27.9 and 30.5cm and was depressed much deeper into the soil than the boot print (made by researcher) seen above the ruler. While the length of the print is not unusual, its depth is highly noteworthy, again indicating that great weight was needed to make the print.

The five photographs that follow show a remarkable footprint find at Buncombe Hollow, Skamania County, Washington in October 1974. Buncombe Hollow is on a narrow dead end road bordering the southern shores of Merwin Dam Reservoir (situated east of Woodland). Scouts, attending a 24-hour watch on slash burning, sensed a "presence" during the night and in the morning saw unusual footprints. They notified Robert Morgan (a noted sasquatch researcher) and he and Eliza Moorman went immediately to the area. They followed the prints, first uphill along the long drag and then down where they entered Buncombe Creek. In all, an unbroken string of 161 prints were counted. The prints were highly unique in that they transversed varying terrain permitting multiple areas to expose toe movements and the compaction of different soil types. Morgan contacted Dr. Grover Krantz who personally investigated the find.

Close-up of a Buncombe Hollow print. It measures about 17-inches/43.2cm long.

Robert Morgan measures prints.

Dr. Tripp's Conclusion on Soil Penetration

In about 1959, an article appeared in the *San Jose News* on findings by Dr. R. Maurice Tripp, a geologist and geophysicist. Tripp went to the scene of a sasquatch sighting in the Bluff Creek, California area and made a cast of a 17-inch/43.2cm footprint he found at the sighting location. He made engineering studies of the soil properties and depth of the footprint.

The following is the photograph and caption that appeared in the newspaper.

He Has Cast As Proof

Dr. R. Maurice Tripp measures a cast of what he says is the footprint of an "abominable snowman." Dr. Tripp says the footprint is that of a man who weighs more than 800-pounds/362.4kg and has been seen by residents of an area near Eureka.

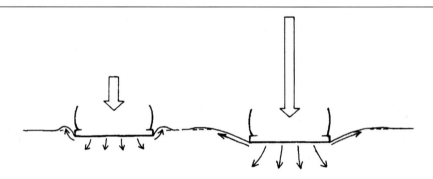

Robert Morgan (left) and Dr. Grover Krantz. Dr. Krantz wrote the following regarding the prints:[1]

"While examining a set of tracks in southwestern Washington with Robert Morgan in 1975 (Morgan says 1974-CLM), the idea of impact faking occurred to me. In this particular instance most of the footprints were in loose dirt, and I had already noticed the pressure mound of dirt that surrounded many of them. A simple experiment showed that when I walked by, a similar pressure mound was pushed up around my own prints. But when I stamped my foot with some force, the dirt was shifted aside with much more speed and no mound developed (Fig. 16). My conclusion was that something there had placed those footprints with upwards of 800 pounds/362.4kg of weight coming down on them with no more impact than from a striding gait."

Figure 16. Pressure mounding. Soil compaction underneath a footprint is a product of impressed weight and speed of impact. These drawings are my interpretation of an experiment with shoes in loose dirt. At walking speed (left), soil is compacted directly under the sole, while some is pushed aside and rises in the direction of least resistance. With more forceful stamping (right), soil compaction is somewhat greater, and the side shifted dirt is moved more rapidly. This rapid movement carries the dirt farther, leaving no mounding and a less distinct foot outline.

Morgan and friends estimate the height of the creature that made the prints.

Morgan demonstrates the creature's pace.

[1] *Bigfoot Sasquatch Evidence*, (Hancock House, 1999) P.42

A straight walking pattern is highly evident in this photograph of footprints found on Blue Creek Mountain, California in 1967. I have been told that some First Nations people walk in this manner.

Alternating pattern of human footprints.

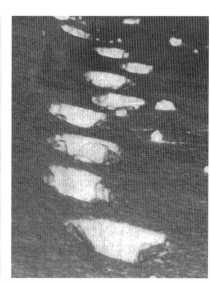

Prints found in different geographical areas (note the straight walking patterns).(Left) Estacada, Oregon, 1968; (Center) Powder Mountain, British Columbia, 1969; (Right) Delox Marsh, Freemont, Wisconsin, 1968.

This illustration shows the foot of the creature seen in the Patterson/Gimlin film and a bear's foot. It will be noted that there is hair between the sole and toes on the bear's foot. It has been reasoned that the dark area between the sole and toes of the creature's foot might also be hair.

NOTE: *This illustration shows the creature's right foot and a bear's left foot. This arrangement was selected so that the toes would match up (i.e., big toes on the inside). However, with bears, the big toe is on the outside (bear foot seen would be reversed if I had used its right foot).*

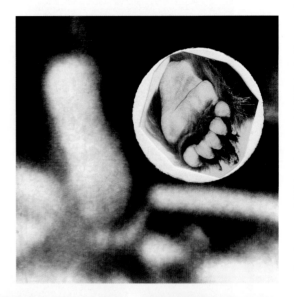

Here, a cast of a double-tracked bear print is compared to the cast of a sasquatch footprint. Double tracked bear prints have been mentioned by some people as a possible reason for "sasquatch" tracks. While double-tracking can add length to a print, it is seen here that there is only marginal similarity between the casts. We can state beyond a doubt that bear prints (double-tracked or otherwise) are definitely not the same as sasquatch prints.

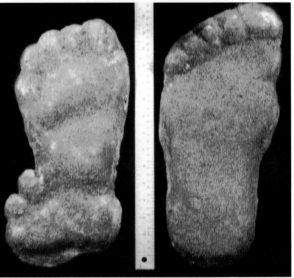

Another example of a double-tracked bear print. This time, the hind foot has landed completely over the print made by the front foot and claws are evident. However, we still see unmistakable bear prints.

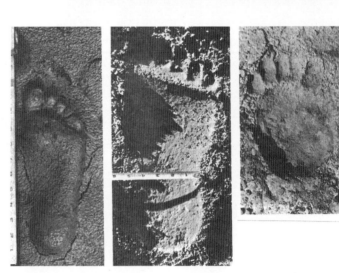

A human, sasquatch and bear (hind foot) print. Bear prints are very different..

Despite the experience, knowledge and good eyesight of numerous people who experience a sasquatch sighting, they are immediately told by authorities that they probably just saw a bear. Sometimes, they are told that they probably saw a hiker with a packsack that made him look much larger at a distance. While we cannot discount these possibilities in some cases, the difference between sasquatch, bears and hikers are very obvious. The following poster by Paul Smith fully illustrates this point.

THE FAHRENBACH FINDINGS

Dr. Henner Fahrenbach, formerly with the Oregon Primate Research Center (now retired) continues to be a major authority on the sasquatch/bigfoot issue. His research spans many decades and he is convinced that there is sufficient evidence to support the existence of the creature. On the question as to why sasquatch credibility is not recognized by the general scientific community he states: *It is easy to put off if you don't know anything about it. However, it is generally uncharacteristic for a scientist to respond in that way. That particular response is reserved for sasquatches.*

I have known Dr. Fahrenbach for nearly ten years and have corresponded with him on numerous issues. He is an extremely meticulous and exacting scientist. He never draws a conclusion unless he is very sure and has either hard evidence or sound statistics to support his stand. The following is Dr. Fahrenbach's findings on his study of sasquatch footprints and other data. (Photograph taken in 2003.)

INTRODUCTION: The data that produces these graphs came predominantly from the records of John Green (Harrison Hot Springs, B.C.), collected over the past nearly 50 years, with additional contributions by J.R. Napier, J.A. Hewkin, P. Byrne, and myself in addition to some details extracted from the Patterson/Gimlin movie. This material was published in extended form in the journal **Cryptozoology** (W. H. Fahrenbach, *Sasquatch Size, Scaling and Statistics*, Vol. 13, 1997-1998, p. 47-75). The raw numerical material was not edited or selected, but used in its entirety. Thereby, the statistical noise was increased somewhat by some spurious data that were presumably included, but no bias was imposed upon them. The area covered includes 10 western U.S. States plus Alaska, and the western Canadian Provinces.

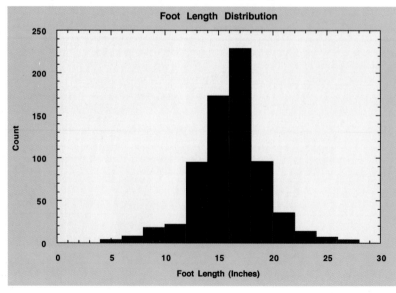

FOOT LENGTH: This histogram comprises 706 footprints, each one of them representing a short or long trackway, the latter sometimes extending over miles. The distribution is bell-shaped, meaning that it came from a biological population rather than being the result of forgery (an approach that would not have yielded the distribution). It is quite peaked, indicating that the males and females of comparable size/age are no more than about a foot different in height (see height graph below). The average foot length is 15.6/39.6cm, the range extends from 4 inches to 27-inches/10.2 to 68.6cm. The average male human foot is about 10.5-inches/26.7cm long.

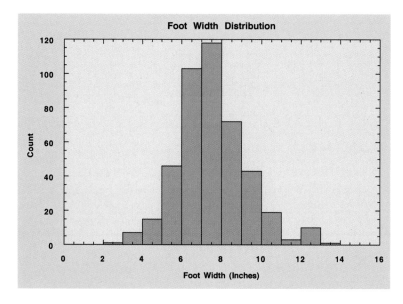

FOOT WIDTH: *This distribution describes the sasquatch foot width at the level of the ball of the foot. The range is 3-inches to 13.5-inches/7.6 to 34.3cm and the average width measures 7.2-inches/18.3cm. Again, the distribution is described by a bell-shaped curve. In this case, 410 footprints were measured for width.*

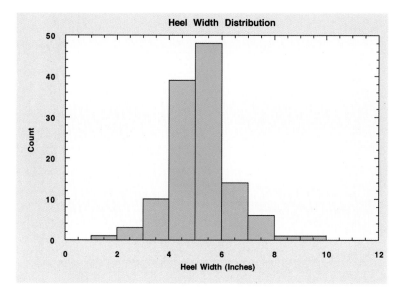

HEEL WIDTH: *Heel width is rarely measured, 117 measurements contributed to this graph. Even this limited sample yields a normal distribution in congruence with foot length and ball width. Heels range from 1.5-inches to 9-inches/3.8 to 22.9cm wide, with the average being 4.8-inches/12.2cm.*

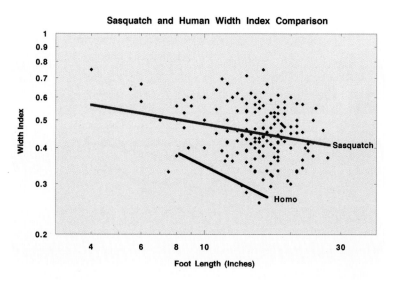

FOOT WIDTH INDEX: *A useful manner of describing the shape of the foot is the width index, meaning the width at the ball divided by the length of the foot. The larger the resulting fraction, the broader the foot is. The upper line, which averages all the data contained in the graph from 410 measurements, hovers about an average slightly under 0.5 with a very slight decrease in relative width with increasing length. By contrast, the lower line indicates the condition in man, in whom the foot gets relatively narrower as its length increases. It appears that sasquatch female feet are narrower than those of males, but insufficient data are at hand.*

STEP LENGTH: *Step length is a much less definable feature in that it ranges from aimless shuffling to full-out running. Usually steps are only measured when they represent a trackway, although even in this context it is often not stated whether the measuring was done from heel to heel or toe to toe rather than just from toe to the next heel. Even if the latter was applied (the wrong way), the result provides a step length that is shorter by the length of the foot. Thus, this graph represents a conservative minimum. Running steps are inherently hard to recover in the usual uneven and duff-covered terrain of the forest.*

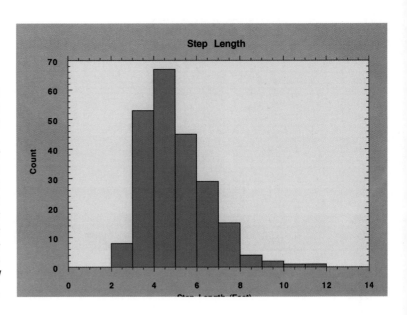

GROWTH: *The growth curve is based on fewer data than any of the preceding graphs, but nonetheless holds some instructive value. Anchor points were provided by the smallest recorded, barely walking feet of infants, arbitrarily designated to be one year old, and at the other extreme those of a few identified female footprints. Three sets of footprints, thought by the respective collectors to belong each to one animal, all collected over a period of years, were fitted between the extremes. Since foot growth, seen here, is different from general bodily growth, the latter would describe a slightly different curve.*

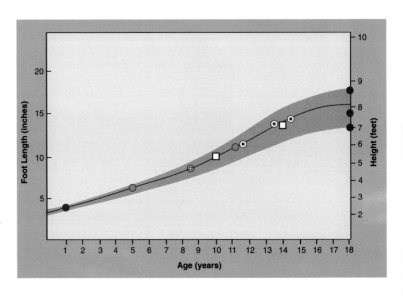

GAIT: *This graph depicts 297 cases in which both foot length and step length was measured. The red line averages all the steps and shows a steady increase in step length with foot length, approximately 5-feet/1.5m for the average-sized sasquatch. The black line is extrapolated (from human walking) to indicate at which level the gait changes from walking to running. Long running steps, though inherently rare in this species, are undoubtedly under-represented due to the difficulty of finding and following them. The approximate speed, based on cadence of 85 steps per minute is indicated in the right Y-axis. The majority of the steps collected here probably came from animals walking at their normal, unhurried pace and were produced in the absence of man.*

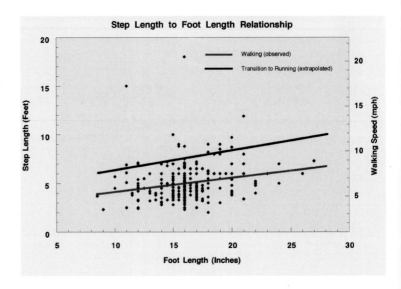

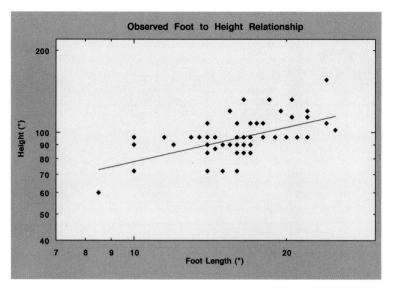

FOOT TO HEIGHT RELATIONSHIP: *In a number of visual encounters, the foot length was measured subsequently and is here plotted against the estimated height. Inspection of the regression line (the average of all data points) shows the surprising detail that for a 20% linear growth of the animal, the foot grows 60%, lending the name "Bigfoot" some statistical credence. The biological reason is to be found in the fact that the weight of the animal rises with the approximate cube of its linear dimensions, thus outstripping the bearing weight of the sole unless the foot grows in excess of the rest of the body. As a consequence, in small animals the foot length has to be multiplied by about 7 to give the height, in average feet by 6, and in large feet by 5.*

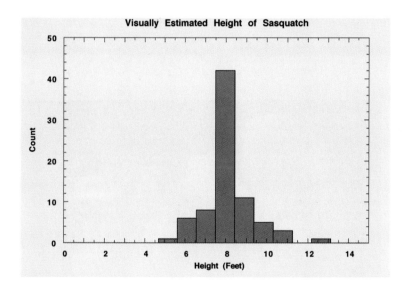

HEIGHT: *Eye witnesses, notoriously inaccurate under the usual circumstances of surprise or fear, account for these records of height estimates. Nonetheless, the distribution is rather evenly centered about an 8-feet/2.4m height.*

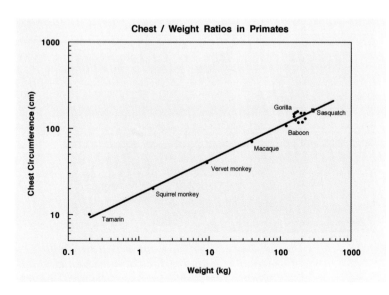

WEIGHT: *Estimates of weight are highly inaccurate, ranging in the case of the Patterson/Gimlin sasquatch through almost a full order of magnitude (280 to 2,028-pounds/125.8 to 906kg). There exists, however, in primates a tight relationship between chest circumference and body weight, ranging from tiny arboreal primates to gorillas. (The gorilla data points represent the weight of individuals, both wild and zoo held, whereas data for other primates are averaged). The chest circumference of the Patterson sasquatch can be derived by geometric means from a picture that includes the full 14.5-inches/36.8cm sole as a yardstick and amounts to 60-inches/152.4cm. That figure entered into the graph yields a weight of 542 lbs/245.5kg. Just like gorillas, sasquatch come in all ranges from skinny to rotund.*

Dan Murphy (left) and Dr. Fahrenbach at the 1995 Sasquatch Symposium in Harrison Hot Springs. Dan is holding the Freeman sasquatch hand cast which Dr. Fahrenbach has diligently studied. There is no doubt in his mind that the cast is from the hand print of an actual sasquatch or bigfoot creature. His conclusions on this cast are found in the Handprints section.

Dr. Fahrenbach talking to reporters at the Willow Creek Bigfoot Symposium, September, 2003.

Part of the speaker's line-up at Willow Creek; Dr. Fahrenbach is on the right. John Green is at left followed by Bob Gimlin and Jimmy Chilcutt.

DR. D. JEFFREY MELDRUM & THE FOOTPRINT FACTS

Dr. D. Jeffrey Meldrum is an Anthropologist with the Department of Biological Sciences, Idaho State University. He has been involved in sasquatch research for over ten years and worked very closely with the late Dr. Grover S. Krantz, Washington State University. He has personally undertaken field research and has seen first-hand what are believed to be sasquatch footprints. He has studied numerous footprint casts, analysed several videos showing what could be sasquatch creatures and has performed a detailed analysis on the Patterson/Gimlin film.

Dr. Meldrum has participated in several television documentaries about the sasquatch, providing highly professional theories and conclusions on the physical aspects of the creature. He lectures on the subject in both the United States and Canada and works closely with the Bigfoot Field Researchers Organization (BFRO). He is the primary professional anthropologist involved in the sasquatch field of studies.

The following presentation is based on posters Dr. Meldrum displays for his talks and lectures. The information provided summarizes all of his findings and conclusions on the subjects to date. The photograph seen here was taken at the Willow Creek, California Sasquatch Symposium held in September 2003. Dr. Meldrum was a keynote speaker at this event.

Evaluation of Alleged Sasquatch Footprints and their Inferred Functional Morphology

INTRODUCTION: Throughout the twentieth century, thousands of eyewitness reports of giant bipedal apes, commonly referred to as "Bigfoot" or "Sasquatch," have emanated from the mountain forests of the western United States and Canada. Hundred of large humanoid footprints have been discovered and many have been photographed or preserved as plaster casts. As incredulous as these reports may seem, the simple fact of the matter remains: the footprints exist and warrant evaluation. A sample of over 100 footprints casts and over 50 photographs of footprints and casts was assembled and examined, as well as several examples of fresh footprints.

Tracks in the Blue Mountains: The author examined fresh footprints first-hand in 1996, near the Umatilla National Forest, outside Walla Walla, Washington. The isolated trackway comprised

"Throughout the twentieth century, thousands of eyewitness reports of giant bipedal apes, commonly referred to as Bigfoot or Sasquatch, have emanated from the mountain forests of the western United States and Canada."

in excess of 40 discernible footprints on a muddy farm road, across a plowed field, and along an irrigation ditch. The footprints measured approximately 35cm/13.75-inches long and 13cm/5.25 inches wide. Step length ranged from 1.0 - 1.3m (39– 50.7-inches.) Limited examples of faint dermatoglyphics were apparent, but deteriorated rapidly under the wet weather conditions. Individual footprints exhibited variations in toe position that are consistent with inferred walking speed and accommodation of irregularities in the substrate. A flat foot was indicated with an elongated heel segment. Seven individual footprints were preserved as casts.

Site of more than 40 tracks

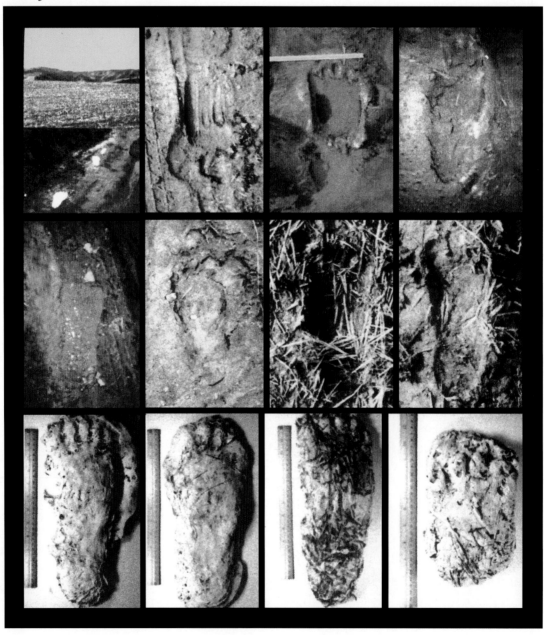

EVIDENCE OF A MIDTARSAL BREAK: Perhaps the most significant observation relating to the trackway was the evidence of a pronounced flexibility in the midtarsal joint. Several examples of midfoot pressure ridges indicated a greater range of flexion at the transverse tarsal joint than permitted in the normal human tarsus. This is especially manifest in the footprint figured below, in which a heel impression is absent. Evidently, the hindfoot was elevated at the time of contact by the midfoot. Due to muddy conditions, the foot slipped backward, as indicated by the toe slide-ins, and a ridge of mud was pushed up behind the midtarsal region.

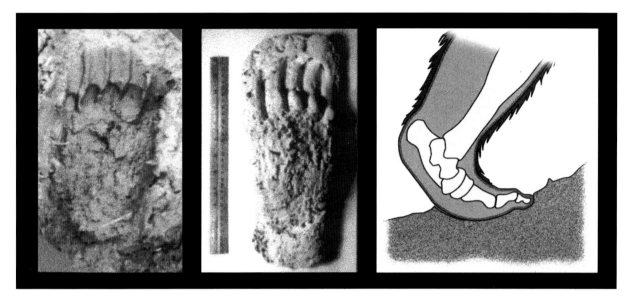

PATTERSON-GIMLIN FILM SUBJECT: In October 1967, Roger Patterson and Bob Gimlin claimed to have captured on film a female Bigfoot retreating across a loamy sandbar on Bluff Creek in Northern California. The film provides a view of the plantar surface of the subject's foot, as well as several unobstructed views of step cycles. In addition to a prominent elongated heel, a midtarsul break is apparent during midstance and considerable flexion of the midtarsus can be seen during the swing phase. The subject left a long series of deeply impressed footprints. Patterson cast single examples of a right and a left footprint. The next day the site was visited by Robert Laverty, a timber management assistant, and his sales crew[1]. He took several photographs including one of a footprint exhibiting a pronounced pressure ridge in the midtarsal region. This same footprint, along with nine others in a series, was cast nine days later by Bob Titmus, a Canadian taxidermist. A model of inferred skeletal anatomy is proposed here to account for the distinctive midtarsal pressure ridge and "half-tracks" in which the heel impression is absent. In this model, the Sasquatch foot lacks a fixed longitudinal arch, but instead exhibits a high degree of

> "A model of inferred skeletal anatomy is proposed here to account for the distinctive midtarsal pressure ridge and "half-tracks" in which the heel impression is absent."

---

[1] Robert Laverty usually goes by his second name, Lyle. Also, his crew was a surveying crew, not a 'sales' crew.

midfoot flexibility at the transverse joint. Following the midtarsal break, a plastic substrate may be pushed up in a pressure ridge as propulsive force is exerted through the midfoot. An increased power arm in the foot lever system is achieved by heel elongation as opposed to arch fixation.

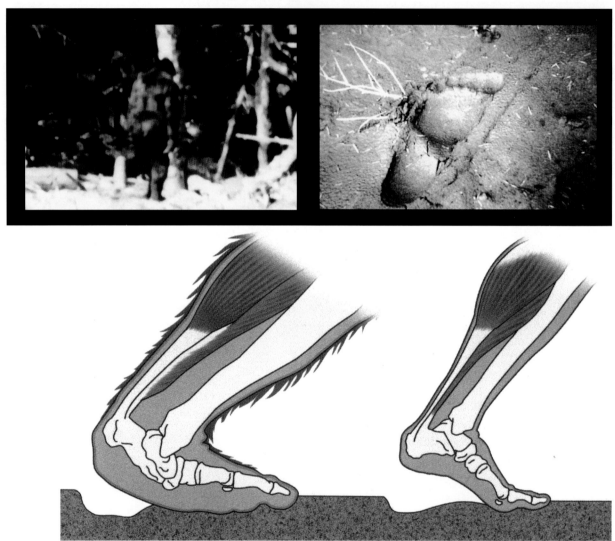

Additional Examples of "Half-Tracks": A number of additional examples of footprints have been identified that exhibit a mid-tarsal break, either as a pronounced midtarsal pressure ridge or as a "half –track" produced by a foot flexed at the transverse tarsal joint. Each of these examples conforms to the predicted relative position of the transverse tarsal joint and elongated heel. The first example is documented by a set of photographs taken by Don Abbott, an anthropologist from the British Columbia Museum, in August 1967. These footprints were part of an extended trackway, comprising over a thousand footprints, along a Blue Creek Mountain road in northern California.

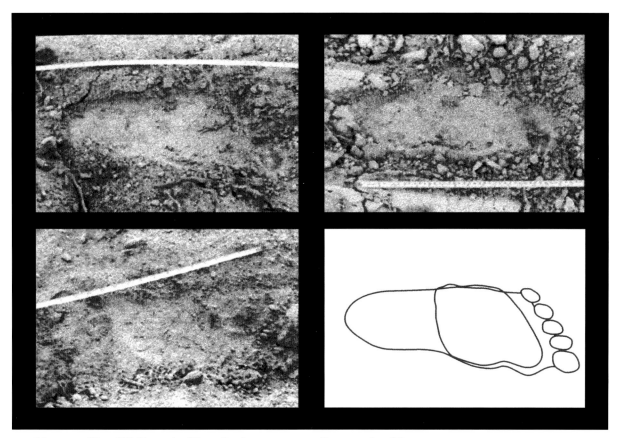

Deputy Sheriff Dennis Heryford was one of several officers investigating footprints found by loggers on the Satsop River, Grays Harbor County, Washington, in April, 1982[1]. The subject strode from the forest across a logging landing, then doubling its stride, left a series of half tracks on its return to the treeline. Note the indications of the fifth metatarsal and calcaneocuboid joint on the lateral margin of the cast. The proximal margin of the half-track approximates the position of the calcaneocuboid joint.

[1] Area is known as Abbott Hill

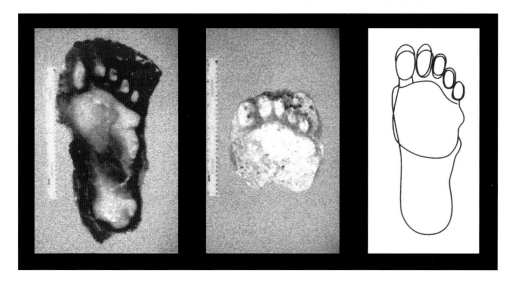

Examples of Foot Pathology: The track of an individual with a presumed crippled foot was discovered in Bossburg, Washington in 1969. The malformed right foot has been previously misidentified as a case of talipes eguinoverus (clubfoot). However, it is consistent with the general condition of pes cavus, specifically metatarsus adductus or possibly skew foot. Its unilateral manifestation makes it more likely that the individual was suffering from a lesion on the spinal cord rather than a congenital deformity. Regardless of the epidemiology, the pathology highlights the evident distinctions of skeletal anatomy. The prominent bunnionettes of the lateral margin of the foot merit the positions of the calcaneocuboid and cuboideometatarsal joints, which are positioned more distal than in a human foot. This accords with the inferred position of the transverse tarsal joint and confirms the elongation of the heel segment. Furthermore, deformities and malalignments of the digits permit inferences about the positions of interphalangeal joints and relative toe lengths, as depicted in the reconstructed skeletal anatomy shown below.

"Its unilateral manifestation makes it more likely that the individual was suffering from a lesion on the spinal cord rather than a congenital deformity."

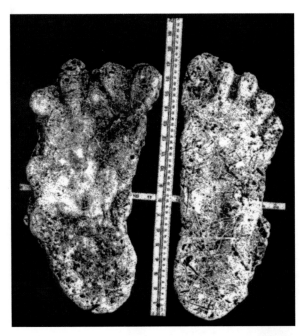
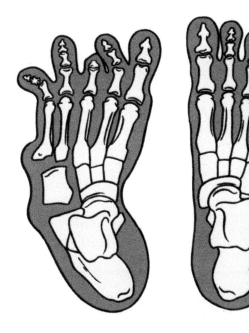

Relative Toe Length and Mobility: Variations in toe position are evident between footprints within a single trackway, as well as between individual subjects. In some instances, the toes are sharply curled, leaving an undisturbed ridge of soil behind toe tips resembling "peas-in-a-pod." In other instances the toes are fully extended. In either case, the toes appear relatively longer than in humans. Among the casts made by the author in 1996 is one in which the toes were splayed, pressing the fifth digits into the sidewalls of the deep imprint, leaving an impression on the profile of marginal toes. This is the first such case that I am aware of.

"...the toes appear relatively longer than in humans."

Expressed as a percent of the combined hindfoot/midfoot, the Sasquatch toes are intermediate in length between those of humans and the reconstructed length of australopithecine toes. Furthermore, the digits frequently display a considerable range of abduction.

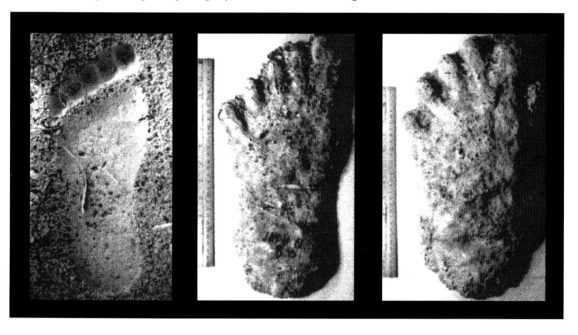

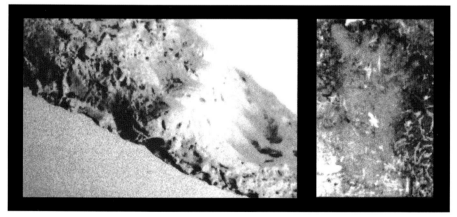

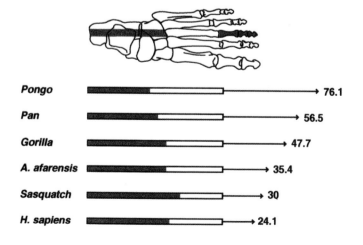

Pongo	76.1
Pan	56.5
Gorilla	47.7
A. afarensis	35.4
Sasquatch	30
H. sapiens	24.1

"...the Sasquatch toes are intermediate in length between those of humans and the reconstructed length of australopithecine toes."

Compliant Gait: The dynamic signature of the footprints concurs with numerous eyewitness accounts noting the smoothness of the gait exhibited by the Sasquatch. For example, one witness state, "it seemed to glide or float as it moved." Absent is the vertical oscillation of the typical stiff-legged human gait. The compliant gait not only reduces peak ground reaction forces, but also avoids concentration of weight over the heel and ball, as well as increases the bound of double support.

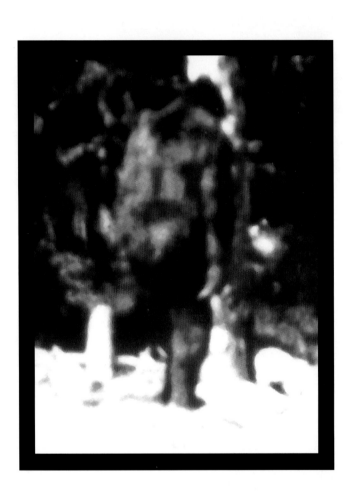

Human walking is characterized by an extended stiff-legged striding gait with distinct heel-strike and toe-off phases. Bending stresses in the digits are held low by selection for relatively short toes that participate in propulsion at the sacrifice of prehension. Efficiency and economy of muscle action during distance walking and running are maximized by reduced mobility in the tarsal joints, a fixed longitudinal arch, elastic storage in the well-developed calcaneal tendon, plantar aponeurosis and deep planter ligaments of the foot.

In contrast, the Sasquatch appear to have adapted to bipedal locomotion by employing a compliant gait on a flat flexible foot. A degree of prehensile capability has been retained in the digits by maintaining the uncoupling of the propulsive function of the hindfoot from the forefoot via the midtarsal break. Digits are spared the peak forces of toe-off due to compliant gait with its extended period of double support. This would be an efficient strategy for negotiating the steep, broken terrain of the dense mountain forests of the Pacific and intermountain west, especially for a bipedal hominoid of considerable body mass. The dynamic signatures of this adaptive pattern of gait are generally evident in the footprints examined in this study.

"In contrast, the Sasquatch appear to have adapted to bipedal locomotion by employing a compliant gait on a flat flexible foot."

Dermetoglyphics in Casts of Alleged North American Ape Footprints

Abstract: Perennial reports of giant apes, commonly referred to as "Bigfoot" or "Sasquatch," have emanated from the mountain forests of the United States and Canada. Hundreds of large humanoid footprints have been discovered. Many of these have been photographed and/or preserved as plaster casts. In some instances, soil conditions were such that dermatoglyphics, or skin ridge details, were preserved in the footprints and transferred to the casts. The casts featured here have been evaluated in collaboration with a professional latent fingerprint examiner. Ridge detail displays distinguishing characteristics including bifurcations, ending ridges, short ridges and scars. However, the dermatoglyphic features are distinct from those of humans in consistent ways. First, the ridges themselves are wider on average than found in humans and non-human primates. Within the hominoids, ridge width is positively correlated with foot size. Second, the pattern of flow of the ridges is distinct. For example, in humans the ridges usually flow transversely across the side of the foot, while in the casts the flow tends to be longitudinal. The possibility of hoaxing is considered and the implications for the existence of an unclassified North American ape are examined.

"Hundreds of large humanoid footprints have been discovered."

Introduction: The presence of dermatoglyphics on Sasquatch footprint casts was first reported at length in the published literature by Professor Grover S. Krantz (1983, 1992). These casts originated from the Blue Mountains of southeastern Washington. This, however, was not the first instance of such skin ridge patterns being noted in footprints. As early as 1967, John Green observed ridges in 33cm (13-inch) tracks discovered on and beside the Blue Creek Mountain road in northern California. The road surface consisted of a very fine rain-dampened dust, "It even appeared to show the texture of the skin on the bottom of the foot, grooved in tiny lines running the length of the print" (Green, *On the Track of the Sasquatch*, [1971], p.47). "It gave the appearance of wood grain; no ridges were noticeable the next day," (Green, personal communication). A cast from this incident preserves definite skin ridge detail, ruling out the possibility of a carved wooden foot.

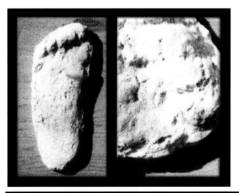

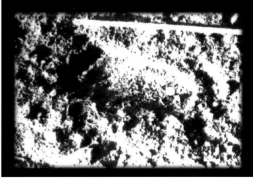

Examination of a set of original casts made by Bob Titmus near Hyampom, California on April 28, 1963, reveals ridge pattern about the digits on one cast, especially the medial side of the hallux where the plantar pad is prominent. Dermal ridges are evident flowing parallel to the edge of the foot. The footprints, measuring over 43cm (17-inches), were found in wet mud. This individual's tracks were found in the region on several occasions over a 6-year period.

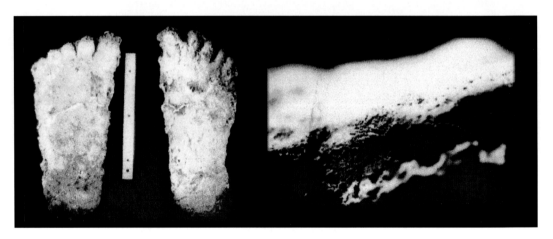

Another instance occurred near Blanchard, Idaho in 1977. Large 42cm (16.5-inch) footprints were discovered crossing a wet, muddy road. Several witnesses observed lines or "veins" in some prints that were interpreted as dermal ridges. These were preserved in subsequent casts. This incident was investigated at considerable length by Dr. James Macleod *et al*, of North Idaho College (personal communication).

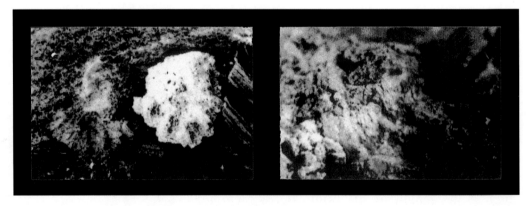

An additional dramatic example of ridge pattern was identified on a set of casts from the Blue Mountains in 1984. Referred to by Krantz as "Wrinkle Foot," these specimens from Table Springs (Walla Walla River) exhibit extensive ridge detail across the plantar surface of the cast. The set includes a right and left pair measuring 33cm (13-inches) long and a partial print of the distal end of the right foot. Krantz noticed that the right and left feet were not precise mirror images of one another and attributed this to geriatric crippling. However, if the midfoot is flexible, as proposed by Meldrum (1999), then the variance can be more readily accounted for by the expression of a greater degree of supination in the right foot than the left. If the right leg was externally rotated, the resulting supination would raise the medial border of the foot slightly, including the proximal end at the hallucial metatarsal and internally rotate the calcaneus relative to the midfoot. These are precisely the distinctions evident in the casts. The left foot is fully pronated and both the distal and the proximal ends of the hallucial metatarsal are fully impressed into the substrate. The impressions of the taut plantar aponeurosis can be seen most evident in the right, more fully pronated foot. The enlarged ends of the joints of the metatarsal permit an appropriate length estimate of 6.5cm (2.6-inches). This is consistent with an evident relatively shorter metatarsophalangeal observed in other casts where flexion creases and bony landmarks permit length estimates. The partial print clearly shows strong dorsiflexion of the metatarsophalangeal joints, confirming the identification and placement of the first metatarsophalangeal joint. This combined with the pronounced plantarflexion of the interphalangeal joints and abduction of the digits, especially the first and fifth. Although the toes conform to the shape of those of the complete right foot, the positions are varied and yet are appropriate to the context of securing a toehold on an inclined bank.

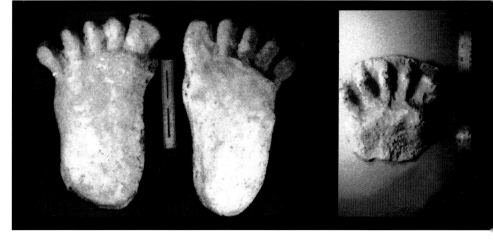

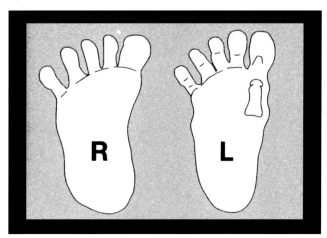

The outline tracing of the left foot is reversed for contrast to the right foot. The inferred position of the hallucial metatarsal is indicated.

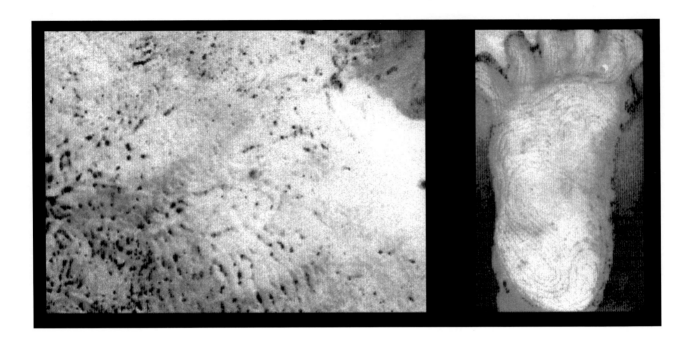

A county Deputy Sheriff in Georgia responded to a repeated disturbance on a farm on the flood plain at the Flint River in 1996. Large 48cm (18-inch) tracks were found on the river bank and extending into the water. A cast was made of one of the clearest tracks by the deputy on duty. Close examination of the cast revealed a dermal ridge pattern of a comparable texture and flow pattern evident in previous casts, which was subsequently confirmed by a latent fingerprint examiner.

"Close examination of the cast revealed a dermal ridge pattern..."

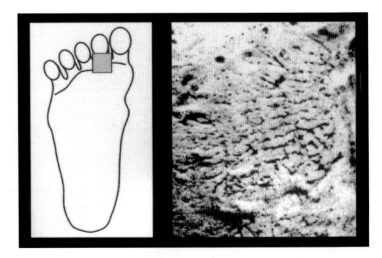

Latent Feature Examination: The casts were examined by Officer J. H. Chilcutt, Latent Fingerprint Examiner, Conroe Police Department, Conroe, Texas. Officer Chilcutt brings not only an expertise in human fingerprint examination, but expertise in non-human primate print examination spanning five years and over

1,000 finger, palm and sole prints. His examination of these casts confirmed the presence of dermal ridge pattern with typical characteristics such as bifurcations, ending ridges, and short ridges. However, the ridge width was on average twice that of human samples and the flow pattern was also distinct. The dermal ridges trend lengthwise along the sole of the foot, especially along the margins, whereas human ridges tend to flow transversely across the sole of the foot. In addition, examples of scarring are present on the Walla Walla casts. Ridge flow interrupted by healed cuts displays characteristic distortions.

Ridge texture, often expressed as ridge count (i.e., ridges/cm) is quite variable among primates, ranging from 35 ridges/cm in some prosimians to 10-15 ridges/cm in Old World monkeys. Even these values can vary within an individual foot. It has been suggested that the dermatoglyphics on the Sasquatch casts could be accounted for by some process such as expanding moulded human dermatoglyphics (Baird, 1989). Baird describes a method of enlarging a latex mold with kerosene. This process was replicated in our lab with latex molds of human feet with clear ridge detail. It resulted in a uniform expansion of the mold and attempts to disproportionately expand selected areas created deformation and warping of the mold. The process also left the mold extremely brittle and difficult to handle without damaging it. More fundamentally, this method fails to address the distinctions of ridge flow pattern evident in the casts.

Conclusion: The existence of multiple, independent examples of footprint casts spanning three decades and thousands of miles, each displaying consistently distinct dermatoglyphics constitutes significant affirmative evidence for the presence of an unrecognized North American ape. The combination of distinctive anatomy of the foot and details of ridge texture and flow make the probability of a hoax unlikely.

"However, the ridge width was on average twice that of human samples and the flow pattern was also distinct."

BIBLIOGRAPHY

Baird, D. (1989), *Sasquatch Footprints: A proposed Method of Fabrication,* (Cryptozoology 8:43-46)

Chilcutt, J. H. (1999), *Dermal Ridge Examination Report* (unpublished)

Green, J. (1969), *On the Track of the Sasquatch* (Cheam Publishing, Agassiz, B.C.)

Horseman, M. (1986), *Bigfoot: The creature has shaken up a lot of people in the area,* (The Tribune, Deer Park, WA, Wednesday, August 6)

Krantz, G.S., Dr. (1983), *Anatomy and Dermatoglyphics of Three Sasquatch Prints* (Cryptozoology, 2:53 -81)

Krantz, G.S., Dr. (1992), *Big Footprints: A Scientific Enquiry into the Reality of Sasquatch*, (Johnson Printing Company, Boulder, Colorado)

Meldrum, D. J., Dr. (1999), *Evaluation of Alleged Sasquatch Footprints and Inferred Functional Morphology*, (American Journal of Physical Anthropology, 28:200)

DERMAL RIDGE PATTERN EXAMPLES

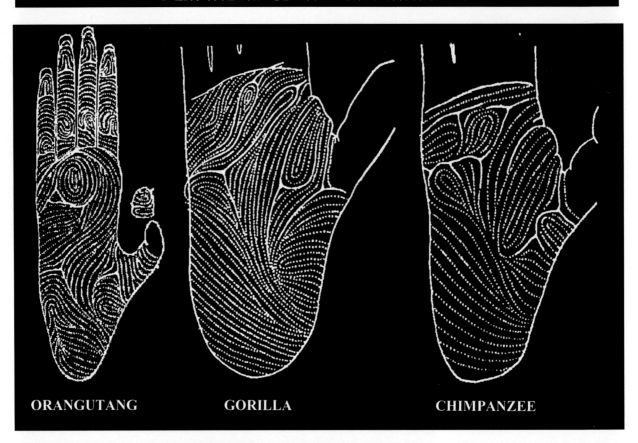

ORANGUTANG   GORILLA   CHIMPANZEE

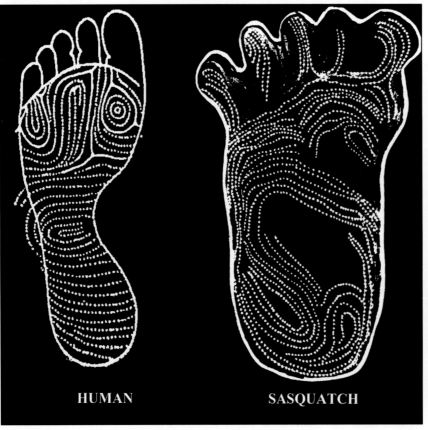

HUMAN   SASQUATCH

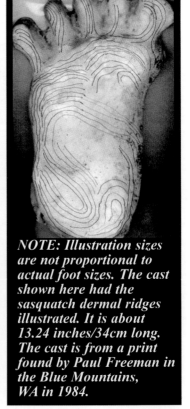

NOTE: Illustration sizes are not proportional to actual foot sizes. The cast shown here had the sasquatch dermal ridges illustrated. It is about 13.24 inches/34cm long. The cast is from a print found by Paul Freeman in the Blue Mountains, WA in 1984.

HANDPRINTS

There is no doubt that sasquatch probably leave handprints as they look for food near rivers and streams. However, the possibility of finding handprints and being able to produce a plaster cast from them is very slim. Footprints alone are rare, but the creature walks on its feet, so there are many situations in which it will make a footprint impression. Nevertheless, alleged sasquatch handprints have been found and highly remarkable casts were produced.

These casts of a knuckle print and handprint were made from impression found by Paul Freeman in Washington State[1]. The human hand shown is that of a large man, about 6-feet (1.8m) tall and weighing about 215 pounds (97.4kg). Concerning the handprint, Dr. Fahrenbach states that the accompanying footprint was 16-inches (40.6 cm) long, which indicates a creature about 7-feet, 4-inches (2.2m) in height. Dr. Fahrenbach concludes: *The fingers may appear shorter and more pointed than they are in reality since sand had started to drift down into the holes left by the fingers. The print is remarkable for the absence of the thenar pad (the bulge at the base of the thumb) and the visibility of the finger tendons within the palm if the cast is held at an angle to a sharp light, both factors indicating a low level of opposability of the thumb. Its width at the palm is also fairly low in comparison to the largest, though less complete prints, that have been found.*

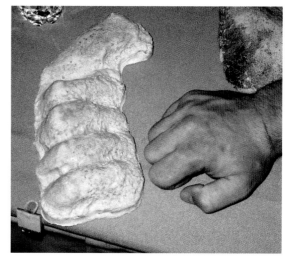

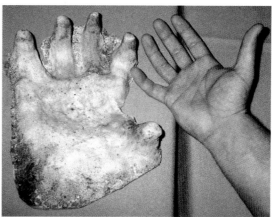

[1] *The knuckle print was found in 1982, Elk Wallow, WA; the handprint was found in 1995, Blue Mountain, WA.*

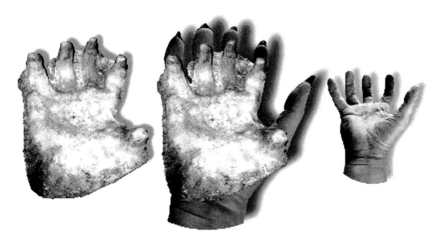

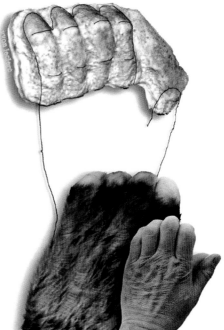

Illustrations by Yvon Leclerc for a better understanding.

Here I provide a shape comparison between a cast of my own hand and the alleged sasquatch hand cast. I photographically enlarged my hand cast to the same size as the sasquatch cast. We can see that the casts are distinctly different.

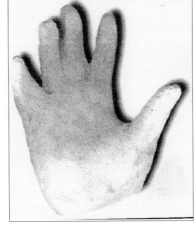
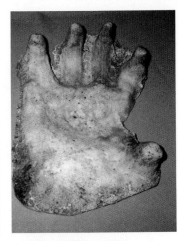

This is a cast of a handprint found by Bob Titmus in the mud at the bottom of a shallow pond, Onion Mountain (Laird Meadow), California area (1982). Titmus drained the pond to make the cast. The length of the print from the tip of the fingers to the end of the palm is about 12-inches/30.5cm.

In February 1962 a sasquatch left a muddy handprint on the side of a white house in Fort Braggs, California. The creature apparently tried to enter the house. The 11.5-inch/29.2cm print was traced and compared with a human hand. While I do not know the size of the man who provided the human print, my own hand is 8.25-inches/21cm long by comparison. I am 6-feet/1.83m tall and have weighed as much as 215-pounds/97.4kg.

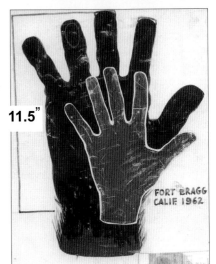

This photograph shows what could be a sasquatch handprint and is more in line with what one would expect to see for such prints. In other words, a flat print, indicating the creature put its hand flat on the ground. This print was found in Ohio (Adams County) in May 1995 by two sasquatch researchers. The print measured about 10-inches/25.4cm from the tip of the longest finger to end of the palm.

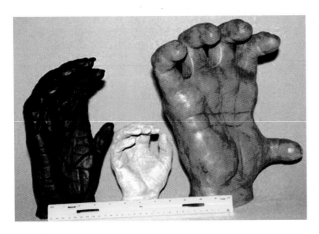

This illustration shows a gorilla hand, human hand and a sasquatch hand created by Caroline Sue Lindley. Sue based the sasquatch hand on the cast made by Paul Freeman previously discussed.

THE BFRO SKOOKUM CAST

In September 2000 Richard Noll, Dr. Leroy Fish, Derek Randles, Thom Powell and other researchers with the Bigfoot Field Researchers Organization (BFRO) were conducting research in the Skookum Meadows area of Gilford Pinchot National Forest, Washington State. On a suggestion by Powell they placed fruit on the ground in an area where there was soft earth and light mud in hopes of attracting a bigfoot creature and obtaining good footprints. When they returned to the area later, some of the fruit was gone and there were a number of known animal footprints and other impressions in the ground. The other impressions indicated that a large animal of some sort had partially laid down in the area and repositioned itself a few times. What could logically be seen as buttocks, thigh, forearm, heel and hand were observed.

Noll reasoned that the impressions could have been made by a bigfoot creature and the other researchers agreed with this possibility. The group thereupon made a large plaster cast of the impressions. The photograph seen above shows Dr. Jeff Meldrum with the cast.

The cast was examined by Dr. Grover Krantz, John Green and Dr. John Bindernagel (seen here, L to R) who concluded that the imprints cannot be attributed to any known animal species. A subsequent examination by Dr. Jeff Meldrum, Dr. George Schaller, Dr. Daris Swindler and Dr. Esteban Sarmiento further confirmed this conclusion. The following drawing by Peter Travers shows the assumed semi-reclining position of the creature as it reached for the fruit. The cast is again shown with the various parts of the creature's body identified.

The multiple impressions of the heel, buttock and thigh were created as the animal repositioned itself at the edge of the mud.

Skookum Cast

HOW THE SKOOKUM PRINTS WERE MADE

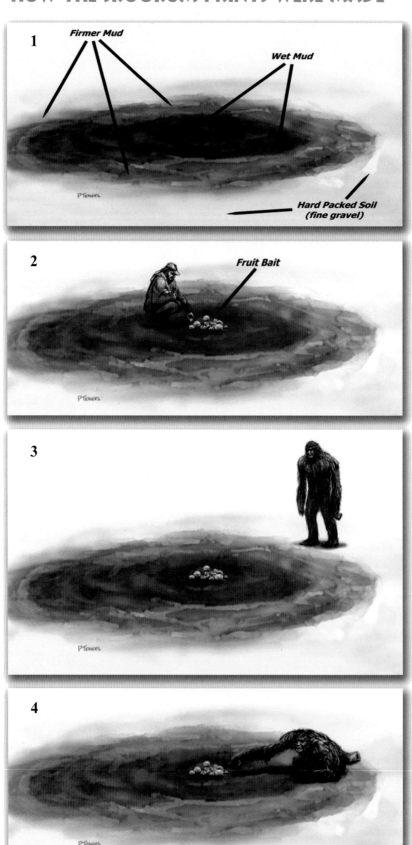

This sequence, created by Pete Travers, shows the layout, preparation and subsequent events that are believed to have resulted in the body prints made by the Skookum sasquatch. The original idea was that the fruit would attract a bigfoot and it would walk into the heavy muddy area and leave good foot impressions. However, the creature that came along chose to stay out of the heavy mud. It laid down in the position shown in the soft earth surrounding the mud and reached in to take the fruit.

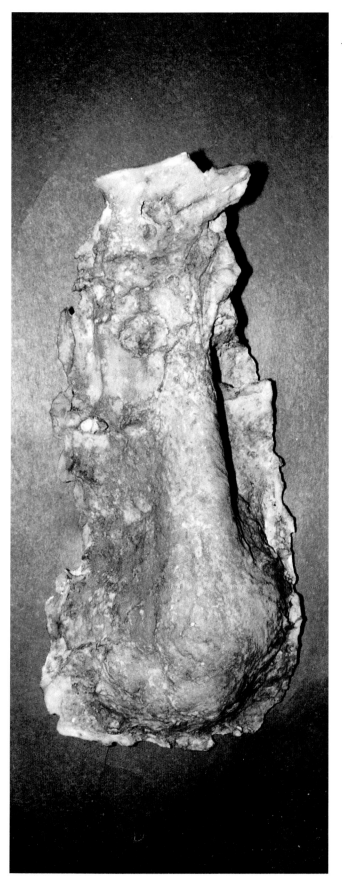

Shown here (left) is a cast made from the creature's heel impression. The heel is much larger than a human heel but definitely appears to be that of a primate. (Note: This cast was made from a mold of the heel.)

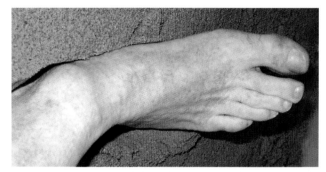

It is reasoned that the creature dug its right foot into the soft soil as illustrated here with a human foot in sand. The angle was such that the impression extended some distance up the back of the leg.

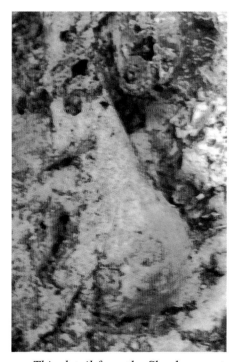

This detail from the Skookum cast shows the actual heel as it appears on the cast.

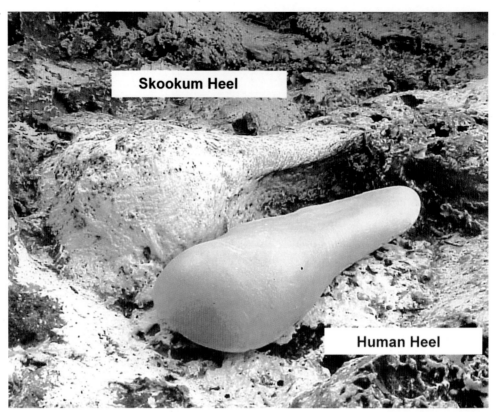

The relative size of the Skookum cast heel is evident here in this comparison with the cast of a human heel.

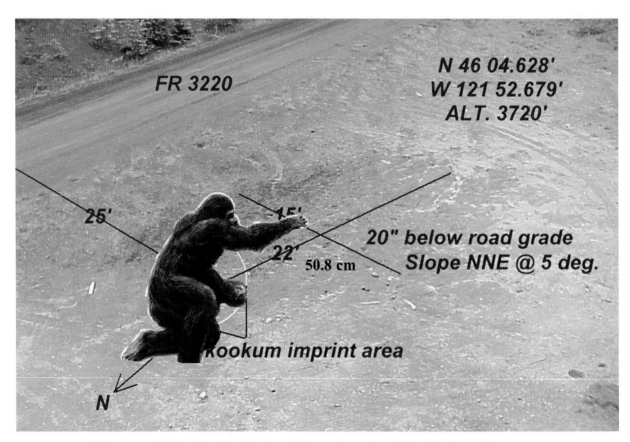

The exact Gilford Pinchot National Forest location of the Skookum imprint area..

150

The Skookum cast under intense examination by (Left to Right) Dr. Jeff Meldrum, Dr. Esteban Sarmiento and Dr. Daris Swindler.

Dr. Daris Swindler is seen here in the 1970s holding casts of gigantopithecus (left) and gorilla mandibles. Dr. Swindler is professor emeritus of physical anthropology at the University of Washington. He is author of the standard text on comparative anatomy of humans and chimpanzees, and has taken an interest in the sasquatch for more than 30 years. He even traveled to Fort Langley to question Albert Ostman in detail about the four creatures he claimed to have seen after one of them carried him off in his sleeping bag. Dr. Swindler remained a skeptic, however, even appearing in documentaries in that role, until he had the opportunity to make a thorough examination of a heel print from the Skookum cast. **He has since stated on camera, without equivocation, that the cast shows the heel of a giant unknown primate.**

Rick Noll, seen here with an assortment of footprint casts, is a primary sasquatch investigator. He played a major role in the Skookum Meadows project and is the custodian of the remarkable Skookum cast. Rick is a man of few words but plenty of action. He is equally at home doing research in the field and sitting at a computer. His contributions to the field of sasquatch studies have been, and continue to be, highly significant.

SASQUATCH HAIR ANALYSIS

With hair analysis, we have a "Catch 22" situation. In order to establish that a hair sample came from a sasquatch, it is necessary to compare the sample with an actual sasquatch hair. If the object of the exercise was to prove the creature exists, it would be redundant because its existence would have already been proven by the actual hair sample. The absolute maximum result we can get from an alleged hair sample is to establish that it did not come from any known creature. The same situation primarily applies to DNA analysis. Here, however, it could be determined if the DNA was in the same order as human beings or other primates. If, for example, the DNA of a wild North American creature was established as being that of a non-recognized primate, then we would have a good case for sasquatch. To my knowledge, we have not been able to get any DNA from hair samples or feces sample that came from alleged sasquatch.

Determining that a hair sample cannot be recognized, however, does allow us to speculate a little. If the sample did not come from something on record, then it definitely came from something not on record. Sasquatch is one creature that is not on record. Unfortunately, there are other animals in this category as hair samples are not on file for every hairy creature in North America or elsewhere.

Remarkably, in 1975 the United States Army Corps of Engineers officially recognized sasquatch as a "possibility" in the *Washington Environmental Atlas.* According to an article in the *Washington Star News* (July 1975) part of this recognition was based on a Federal Bureau of Investigation (FBI) analysis of a hair sample that could not be found to belong to any known animal.

The entry in the atlas is shown on the next page with the accompanying text reprinted on the right. The chart shown on the entry is reprinted below.

> "If the sample did not come from something on record, then it definitely came from something not on record."

Washington Environmental Atlas

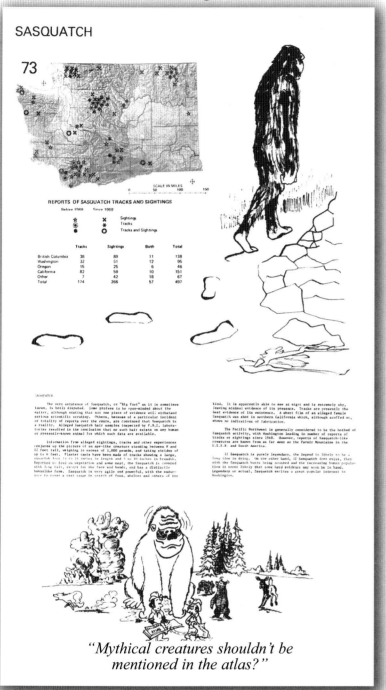

"Mythical creatures shouldn't be mentioned in the atlas?"

	TRACKS	SIGHTING	BOTH	TOTAL
British Columbia	38	89	11	138
Washington	32	51	12	95
Oregon	15	25	6	46
California	82	59	10	151
Other	7	42	18	57
TOTAL	174	266	57	497

SASQUATCH: The very existence of Sasquatch, or "Big Foot" as it is sometimes known, is hotly disputed. Some profess to be open minded about the matter, although stating that not one piece of evidence will withstand serious scientific scrutiny. Others, because of a particular incident or totality of reports over the years, are convinced that Sasquatch is a realty. Alleged Sasquatch hair samples inspected by F.B.I. laboratories resulted in the conclusion that no such hair exists on any human or presently known animal for which such data are available.

Information from alleged sightings, tracks and other experiences conjures up the picture of an ape-like creature standing between 8 and 12 feet (2.4 to 3.7m) tall, weighing in excess of 1000 pounds (453 kg) and taking strides of up to 6 feet (1.8m). Plaster casts have been made of tracks showing a large, squarish foot 14 to 24 inches (35.6 to 60.9cm) in length and 5 to 10 inches (12.7 to 25.4cm) in breadth. Reported to feed on vegetation and some meat, the Sasquatch is covered with long hair, except for the face and hands, and has a distinctly humanlike form. Sasquatch is very agile and powerful, with the endurance to cover a vast range in search of food, shelter and others of its kind. It is apparently able to see at night and is extremely shy, leaving minimal evidence of its presence. Tracks are presently the best evidence of its existence. A short film of an alleged female Sasquatch was shot in northern California which, although scoffed at, shows no indication of fabrication.

The Pacific Northwest is generally considered to be the hotbed of Sasquatch activity, with Washington leading in number of reports of tracks or sightings since 1968. However, reports of Sasquatch-like creatures are known from as far away as the Parmir Mountains in the USSR and South America.

If Sasquatch is purely legendary, the legend is likely to be a long time in dying. On the other hand, if Sasquatch does exist, then with the Sasquatch hunts being mounted, and the increasing human population it seems likely that some hard evidence may soon be in hand. Legendary or actual, Sasquatch excites a great popular interest in Washington.

The last piece of information in the newspaper article, that concerning analysis of hair, prompted considerable interest among sasquatch enthusiasts. An enquiry asking for confirmation and specifics on the analysis received the following response from the FBI:

Since the publication of the "Washington Environmental Atlas" in 1975, which referred to such examinations, we have received several inquiries similar to yours. However, we have been unable to locate any references to such examinations in our files.

The FBI did, however, follow-up with a Dr. Steve Rice, who was editor of the Army Atlas. In an official report, the FBI states: *After checking, Dr. Rice was unable to locate his source of the reported FBI hair examination.*

Joedy Cook

The foregoing information on the Army Atlas and other information relative to the FBI was obtained by Joedy Cook under the Freedom of Information-Privacy Act. Joedy wrote to the FBI and requested all information relative to bigfoot. He received documentation that primarily deals with the Army Atlas and the enquiry previously discussed. There was nothing else of any significance in the file. However, it appears there is always a little "mystery" whenever the FBI is involved in anything. Attached to the file sent to Joedy Cook was a standard pre-printed "Dear Requester" form. Curiously, a box on this form which states, "See additional information which follows," is checked. At the bottom of the form the following information was manually typed in.

> Enclosed are previously processed documents which relate to "Big Foot." The enclosed are the best copies available. Serial 4 is missing from file 95-213013, the file where your release originates. Our effort to locate that document was not successful. It is possible that the number 4 was missed during the original serialization of the file.

George Clappison

We are left to wonder what was in Serial 4. Certainly the FBI should be a little more efficient in their filing procedures. We have, however learned that analysis of hair samples as indicated in the *Washington Star-News* article *definitely took place.* Also, that the samples could not be identified. George Clappison did extensive research on this incident and was referred by the FBI to the ex-head of their Hair and Fiber Unit. This person, who now operates his own private laboratory out of his home, was in charge at the time the hair samples were submitted to the FBI. He told Clappison that *the analysis was done after hours on employee's own time with the results as indicated.* He further stated that no written reports were prepared on the analysis. In discussing the whole situation with the current head of the FBI Hair and Fiber Unit, Clappison asked if the unit would now consider analyzing other hair samples. The current manager agreed to perform an analysis, however, he informed the unit would not respond in writing on their findings.

This is the only case known to me in which bigfoot evidence resulted in some movement toward actual U.S. Federal Government recognition of the creature.

CURRENT RESEARCH: Dr. Henner Fahrenbach has analyzed allged sasquatch hair. He tells us that such hair very closely resembles human hair. The main differences he has established are as follows:

1. Human hair of a 4-inch/10.2cm length always has a cut end unless it originates from the head of an infant who has not yet had a first haircut. The possible sasquatch hair always has native terminations (uncut ends that are worn, rounded or split).

2. Suspected sasquatch hair, whether a straight or wavy shaft, has the same round to oval cross-section for its entire length, and has no color banding. Shorter hairs are likely to have some taper toward the tip.

3. Color of suspected sasquatch hair, when viewed under the microscope, always includes a red tinge plus a vairable amount of very fine pigmentation (melanin) granules. No matter whether the hair looks black, brown or red to the naked eye, it shows the reddish tinge under high magnification.

In my opinion, Dr. Fahrenbach's first observation is truly remarkable. Other than the exception he mentions, it is difficult to justify how the hairs would have two "native terminations" if they were simply human hairs. That they could have been purposely "broken off," to eliminate a cut end must, of course, be considered. Here, however, we have another one of those situations where the probability is far-reaching. Dr. Fahrenbach summarizes his position on hair analysis as follows:

Forensic hair analysis is always based on A/B comparison, which is something we cannot do for obvious reasons. But I am trying to get close by the statistics inherent in the collection of an increasingly sizeable set of hair samples, whose congruity becomes less likely to be due to chance.

The following hair comparison presentation was provided by Dr. Fahrenbach.

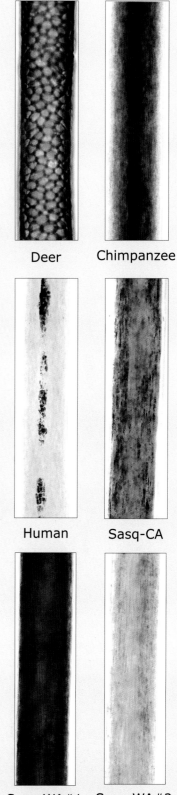

Deer   Chimpanzee

Human   Sasq-CA

Sasq-WA#1   Sasq-WA#2

Hair Micrographs (260x): The deer hair has the cross-section almost entirely occupied by the medulla, an unbroken lattice in hair terminology. It has, of course, a thin cortex and cuticle. The chimpanzee hair, pitch black, has a continuous, mostly amorphous medulla. The human hair has the typical amorphous fragmentary medulla. The three sasquatch hairs (one from California; two from Washington) are of the common color variants – reddish brown (called buckskin by the observers); very dark, probably appearing black; and very dark with a more brown tone, but still appearing black in the field. None of these hairs has a discernable medulla.

SASQUATCH BEDS, NESTS, BOWERS OR HOLLOWS

We can reasonably assume that sasquatch make some kind of bed or "nest" to protect themselves from dampness when they lay down to sleep or rest. Rough structures made of forest material which have been call "bigfoot beds" have been found. It has been observed that "bear grass" has been used to line some of the beds. This type of grass is very sharp and difficult to pull out. To remove it from the ground, a creature would need a "hand," such as a squirrel or raccoon has, as opposed to a paw. The beds are far too large for these types of creatures, so the obvious conclusion is either a human being or something like a human being made the structures.

The following remarkable account on the discovery of a possible sasquatch nest is from Robert J. Alley's book, *Raincoast Sasquatch*. (Hancock House, 2003). The account was provided to Alley by Eric Muench, a Ketchikan, Alaska timber cruiser and logging engineer. Mr. Muench reported:

I had been on Prince of Wales Island working as an independent timber cruiser and logging engineer. On January 26, 1988, on a job for (a local Native corporation) on their land, I was on a hillside above Klawock Lake doing timber reconnaissance to plan some logging units for their coming season. It had been a fairly open winter, and there had been less than one foot of snow under the western hemlock and western red cedar forest at the five hundred foot elevation.

I noticed a patch of huckleberry bushes on the hillside below me that had been broken off uniformly at the four or five foot height. Looking closer, I found a large nest of crudely woven huckleberry branches and cedar bark strips and boughs, lined with mosses and more bark. The circular nest was about seven and one half feet on the outside with a four and one half foot diameter hollow part inside. It was uncovered, but well-placed on the lower side of a down hill leaning red cedar with lots of live feathery boughs hanging directly over the nest, like a natural shingle roof. It was on about a ten-foot wide gentle bench, beyond which a series of small cliffs dropped on down the hill. Nail or claw marks on the tree showed where material had been gathered, and the surrounding ground was stripped of grasses also. The site was less than one-quarter mile above the Klawock Hollis Highway.

Close-up of Kiawock Lake nest with ax for scale.

Distant view of the nest.

In my experience, most bears hibernate in a convenient windfall den, hollow tree or similar partial shelter, with little or no preparation or housekeeping. I have also seen where mother bears will pull in moss, grass or brush tips, probably to warm and soften the place a bit for their cubs. However, this was quite different. Not only were the nest materials somewhat woven together in a way that no bear could do, but the huckleberry bushes had been broken off cleanly, as though two hands had bent the stems so sharp that they could not splinter.

Large cedar beside Kiawock Lake nest showing bark stripped 12-feet/3.7m up.

I wandered around the area a while to look for tracks on deer trails and passages through the cliffs, but the snow was mostly fresh from that day and still falling, so I found nothing. I did pull some fairly stiff, long and slightly kinky black hair from the nest and saw what appeared to be a louse egg on one. It reminded me of horse mane hair, not bear or wolf. The scratch marks, to about six or seven feet up from the ground, clearly showed individual hand pulls. The scratch spread was about eight inches, similar to my own fingers if I spread them way out, but at that spread I could not put scratch-making pressure on my thumb and little finger. I tried, and could not begin to match those marks.

Parallel 8-inch/20.3cm span marks in cedar where bark had been stripped.

While continuing logging road location work the next day, I visited the site again. It had not been disturbed. I designed the logging layout so that the immediate area of the nest was included in a timbered leave strip that protected a deep gorge nearby.

Because I had recently read a down-south 'bigfoot searcher' declare that he intended to prove their existence by offering a reward for a shot specimen, I was reluctant to spread word of my find and risk 'outside' [non-Alaskan] clowns crawling all over my client's property. However, I decided that two people had a right to know. [Mr. Muench named a former land manager for the Native corporation and the logging superintendent for the privately contracted logging company.] I knew them both to be honest, intelligent and thoughtful men and had no hesitation in letting them decide how far to spread word of the nest. Both took the news calmly and without skepticism. In the following days I heard accounts of frequent past sasquatch encounters, including both the Tlingit and Haida names for them, mostly from Native people who had grown up in the Craig and Klawock area. Apparently, knowledge of and belief in bigfoot is common in the area but not often spoken of to strangers from outside the area.

"Apparently, knowledge of and belief in bigfoot is common in the area but not often spoken of to strangers from outside the area."

On February 9, during a Forest Practice Act inspection, the land manager and I took an Alaska Division of Forestry forester and an Alaska Fish and Game habitat biologist to the nest. The biologist gathered a sample from some unfamiliar (at least to me) small dropping piles.

Later that spring or summer, I returned with a camera to photograph the nest and scratch marks, etc., using my ax and a six-inch ruler, for relative scale on the pictures. By that time the nearby brush had '"leafed out" and the boughs in the nest, originally green, had turned brown.

> "I heard a series of slow, measured raps, as though a heavy wood chunk was being swung against a tree."

My only other observation of anything unusual in the area was that, on several occasions during that time on that hillside, I heard a series of slow, measured raps, as though a heavy wood chunk was being swung against a tree. I work alone, and knew that there was no other person anywhere near the area. Following in the apparent direction of the sounds never revealed anything. Years later I was told that such rapping has often been associated with bigfoot sightings or evidence.

I am aware that other people in various capacities visited the nest afterward and before it became destroyed by 'wind-throw' and fire. However, none of that was part of my experience.

A possible sasquatch nest seen here was found in Ohio in 1995. The nest was made of loosely arranged forest material (forming a circle). Close inspection of the nest interior revealed small twigs, branches and dead grass, which appeared "pushed down" or compressed.

We can also reason that sasquatch possibly construct some sort of rough shelters - which more appropriately might be called "bowers" or "hollows." Three unusual structures seen in the following photographs were found in Ohio. Sasquatch researcher George Clappison is seen inspecting the structures. *Certainly, there are other explanations for such structures and I make no claim that what is seen here is even remotely connected with sasquatch. However, as with many things associated with the creature, we can't totally rule out the possibility.*

The first "hollow" or "bower" photograph (which shows the first structure discovered) created a little excitement in 1995 upon being shown on a television program. The structure became known as the Ohio Bigfoot Nest. The specifics are as

Sasquatch nest.

follows. Researchers Joedy Cook, George Clappison and Terry Enders found the structure in Kenmore, a suburb of Akron, Ohio. They had gone to Akron to interview a father and son who had reported unusual, possibly sasquatch related occurrences in the Kenmore area. During the team's investigation they happened across the structure. It was not located in what one might refer to as "a remote area." Although it was in a vacant sparsely wooded section, there is considerable development nearby in most directions. The land was privately owned and part of it had been used as a dumping ground for construction waste and other debris. Nevertheless, there was access to a fairly large forested region, although beyond that there was further development. The structure was hollow on the inside with enough room to accommodate three men in a seated position. It appears to have been made by using large tree branches to form a tunnel. Smaller branches were placed on top followed by vines and weeds with a final covering of long grass.

One can reason that with a covering of snow the structure shown would be a fairly snug retreat, much like an igloo. The research team later found the other similar structures shown (last two photos) in different Ohio areas.

I have been informed by a resident of Akron, John Sawvel (who provided detailed maps), that the Akron "nest" is quite close to a large building. This fact would indicate that any connection with sasquatch would be very remote. However, the researchers were told by the 43-year-old father that he has been aware of an unusual creature in the area since he was a boy. Both father and son stated they had "glimpsed" what they believed was the creature. They also related a story whereby somebody or something in the area threw large rocks at them from a great distance. I really have no way of rationalizing any of this information. Like many sasquatch related reports the alternatives (hoax, imagination, hallucination, whatever) are just as remote as the testimony.

There are some reports that indicate sasquatch live in caves. Most of us have less trouble with this possibility than with bowers or hollows. At this time, however, everything on this aspect of the creature's existence is pure speculation.

In the late 1990s I authored a book with Joedy Cook and George Clappison on bigfoot sightings and other evidence in Ohio. The amount of highly credible evidence in that state is truly remarkable. The book was published in 1997.

Sasquatch "hollows"

Joedy Cook, 1997

SASQUATCH SOUNDS

There are three types of sounds that may be attributed to sasquatch. First, and most intriguing are actual vocalizations that have been taped by a number of researchers. The most noteworthy are the "Sierra sounds" that were originally recorded by Ron Morehead and his company of regular hunters and others in the Sierra Mountains, California between 1971 and 1976. Ron is shown here with a cast of a footprint found near the group's camp in 1972. After the group recorded some sounds in that year, Al Berry, a reporter for the *Redding–Record Search Light* newspaper was invited to come to the campsite. Berry interviewed all of the researchers and heard and recorded unusual sounds himself. He later wrote a full account of the incident and arranged for a professional analysis of the recordings.

The unusual growl-like sounds and whistles were studied by Dr. R. Lynn Kirlin, a professor of Electrical Engineering at the University of Wyoming. It was his opinion that that the format frequencies found were clearly lower than for human data and their distribution does not indicate they were the product of human vocalizations and tape speed alteration. Further, Nancy Logan, a linguist in California also studied the tapes. In her opinion the vocalizations have a pitch range that is considerably more flexible than that of humans. They go much lower and much higher. In Nancy Logan's own words, *The rapid articulation is virtually impossible for a man to do. The vocalizations seem to have some element of language, i.e., certain repeated phoneme patterns and a certain organization to the chattering...they are trying to communicate.* The entire story of how the sounds were obtained and the sounds themselves were skillfully arranged on a tape (later a CD), narrated by Jonathan Frakes, and made available for all sasquatch researchers. Ron Morehead has since recorded more sasquatch vocalizations and has put out a new CD.

Other unusual sounds are those of rocks beings hit together (sometimes rhythmically) and stumps or logs being beaten with a thick stick or tree branch. It is difficult to attribute such sounds to other forest creatures because, it is reasoned, that hands are needed to hold the rocks or stick. Ron's new CD contains some of these sounds together with sounds of tree limbs being snapped (broken branches in remote forest areas have been attributed to sasquatch for many years). He points out that from his experience, sasquatch can mimic many forest animals. While sound cannot be considered hard evidence as to sasquatch existence, they are nonetheless evidence and provide further insights on the possible nature of the creature.

Ron Morehead

"...sasquatch can mimic many forest animals."

SASQUATCH SUSTENANCE

The question as how sasquatch get enough food to sustain themselves is interesting. Here we have what is evidently a large primate living in a very cold, wet and hostile environment. We can certainly point to other large animals (moose, elk, bears etc.) that have no trouble obtaining food, however, these animals are not primates, and are naturally equipped, as it were, to live in North America. Nevertheless, some food highly suitable for sasquatch is naturally available for all or part of the year. Fish, berries, roots and other vegetables fall into this category. Further, if we are to believe some sighting reports, then sasquatch hunt and kill deer for food. Also, after Europeans arrived in North America, cultivated food and domestic livestock would definitely be a food source - there are many sighting reports involving farmers. Further, although not a major food source, unattended campsites and garbage bins or dumps are also on the list.

Seen here is the pit dug into the rocks by the male sasquatch foraging for food with his family. Jim Green, the young man standing in the pit, is 5-feet, 10-inches/1.78m tall. The rock stacks are seen in the background.

In a discussion with Frank Beebe, formerly of the British Columbia Provincial Museum, Beebe informed me that he has absolutely no concerns as to the creature's ability to sustain itself. He pointed out that Arctic grizzly bear sustenance is infinitely less probable and the creature exists. This creature has only four months to find enough food to last it an entire year. Beebe firmly concluded that sasquatch would be able to find more than adequate food sources all year round.

Remarkably, there is very strong evidence that suggests the creatures actively forage for rodents. Three sasquatch, a male, female and juvenile, were observed in November 1967 by Glen Thomas in a natural rock and bolder pile, near Estacada, Oregon, searching for rodents. When rodents were found and caught the sasquatch

Glen Thomas

A distant view of the natural rock pile near Estacada, Oregon in which Glen Thomas observed sasquatch foraging for rodents (probably ground squirrels).

On the left is an elevated view of the hole dug by the sasquatch. Below is a close-up view of the rock piles.

would eat them. The process was unusual. The adult sasquatch would lift rocks, smell them and then stack the rocks in piles. If the smell of the rock indicated the presence of a rodent (logical assumption), then the sasquatch would dig furiously to find their prey. Just why the rocks were stacked is not known, possibly it was to indicate that each rock had been "processed." The largest sasquatch, assumed to be the adult male, actually dug a deep pit in the rocks and removed nesting material believed to contain dormant (hibernating) ground squirrels.

Certainly, the most authoritative and exhaustive research undertaken on sasquatch sustenance (and all other aspects of the creature's life and habits) is that undertaken by Dr. John Bindernagel, a wildlife biologist, who lives on Vancouver Island, British Columbia. Dr. Bindernagel, provides all of his findings and opinions in his book, *North America's Great Ape; The Sasquatch* (Beachcomber Books, 1998). The following is Dr. Bindernagel's Introduction to his book.

"...sasquatch appearance and behavior, although sometimes disturbingly humanlike, resemble most closely patterns of appearance and behavior described for the great apes of Africa and Asia."

My goals in this book are threefold. First I wish to bring together much of what is known - or at least reported - about sasquatch appearance, sign, food habits, and behavior. Many people are unaware of just how many reports of sasquatches or sasquatch tracks exist, for how long they have been reported, and over how large a geographic area they occur. There may also be many people who don't realize that sasquatches have been observed not just striding away, but actually doing things such as digging in the ground, brandishing sticks - even throwing rocks.

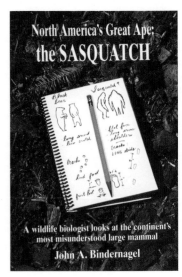

Further I wish to draw attention to patterns evident in these reports and a remarkable consistency in physical features and behavior. This consistency occurs despite the huge time span of over 150 years, the vast geographical area covering most of the states and provinces of the United States and Canada, and the diverse cultural background of eyewitnesses, from trappers in remote locations to police officers on the edge of urban areas.

Lastly, I wish to show that most aspects of sasquatch appearance and behavior, although sometimes disturbingly humanlike, resemble most closely patterns of appearance and behavior described for the great apes of Africa and Asia - the chimpanzee, the gorilla, and the orangutan.

In the end you may conclude, as I have, that the sasquatch is a very real wildlife species, and is, indeed, North America's great ape.

"In the end you may conclude, as I have, that the sasquatch is a very real wildlife species, and is, indeed, North America's great ape."

Dr. John Bindernagel is a wildlife biologist with over thirty years of field experience. He has served as a wildlife advisor for United Nations projects in East Africa, Iran, the Caribbean and Belize. His interest in the sasquatch dates from 1963, and his field work in British Columbia began in 1975. He holds a B.S.A. from the University of Guelph, and an M.S. and Ph.D from the University of Wisconsin. He continues to work as a consultant in environmental impact assessment and is a Registered Professional Biologist (R.P.Bio) in British Columbia, Canada.

Food scarcity, of course, would be a problem for sasquatch as well as all other animals. However, one animal, the wolverine (seen here), has worked out a way to alleviate the problem. This creature takes food (meat) up mountains and buries it in snow. When the going gets tough, the wolverine retrieves its frozen stash and takes it down to a lower elevation to thaw and eat. It has been reasoned that sasquatch might also use this method to ensure an on-going food supply. Interestingly, we might conclude that this is the reason the yeti is observed at such high elevations.

SASQUATCH SPECULATIONS

Many people who have seen the creature remark that the creature in the Patterson/Gimlin film is the same or very close in appearance to the creature they saw. Others, however, are not as positive. Nevertheless, using the shape of the creature's head as seen in the film, together with other indicators, Yvon Leclerc has come up with some reasonable speculations on the creature. The adjacent illustration shows Yvon's significantly enhanced profile of the creature as seen in frame 339 in comparison with different types of primate skulls – a modern human being, two types of prehistoric humans and a female gorilla.

FILM ENHANCEMENT

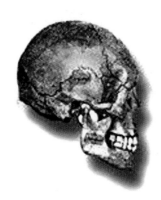 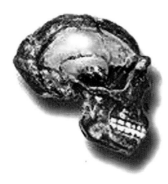  

HUMAN BEING   **PITHECANTHROPUS**   **NEANDERTHAL**   **FEMALE GORILLA**

When the skulls are superimposed onto the profile, we see the following:

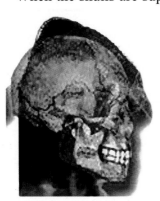 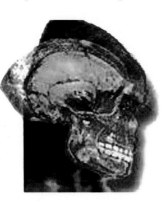  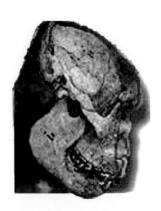

HUMAN BEING   **PITHECANTHROPUS**   **NEANDERTHAL**   **FEMALE GORILLA**

It is seen that a very good match is made with the lower skull (jaws) of the Pithecanthropus and the upper skull (cranium) of the female gorilla.

A composite skull (upper female gorilla; lower pithecanthropus) is superimposed here revealing a remarkable match. This finding just might indicate we are dealing with a creature that is truly an "*ape-man*," or "*man-ape*," a conclusion reached and published by the Russian hominologist, Dmitri Bayanov, in the 1970s.

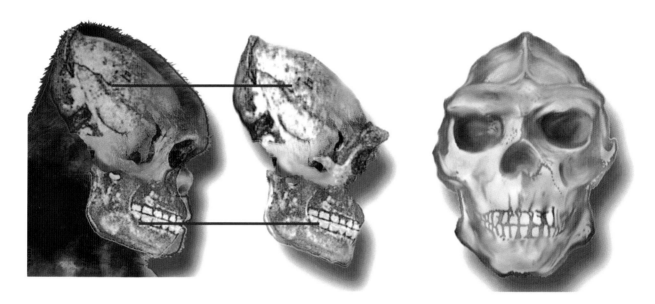

Subsequent research carried out by Yvon substantiated his finding. Using a different film frame (frame 343), Yvon defined the head image and again superimposed the composite skull. The following illustration shows the results. The image on the left is the actual film frame image.

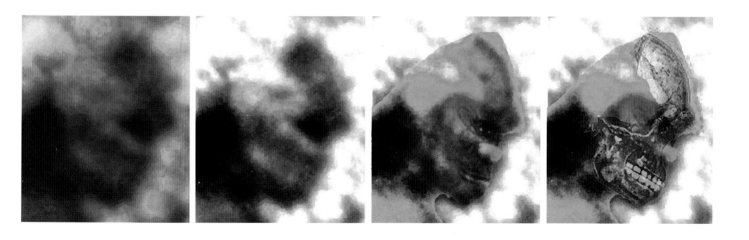

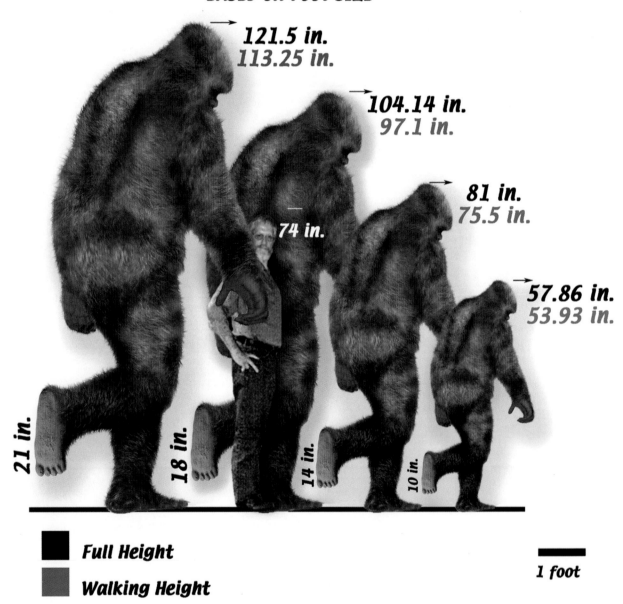

Using the creature in the Patterson/Gimlin film as a model, this illustration by Yvon provides some insights into sasquatch heights and proportions in comparison to a human being. The creature's height has been calculated in accordance with its foot size.

SASQUATCH SIGHTING AND TRACK REPORTS

The following maps show sasquatch sighting and track reports in the western provinces and states indicated up to 1980. The charts were prepared by John Green from his own records at that time. There have, of course, been many additional reports over the past twenty-four years, so the charts are not up-to-date. Nevertheless, they provide a good appreciation of the tremendous range over which the creature has been seen or has left its footprints. (From John Green, *The Best of Sasquatch/Bigfoot,* Hancock House Publishing, 2004)

British Columbia
Sighting and track reports in Britiish Columbia

Washington and Oregon

Sighting and track reports in Washington and Oregon

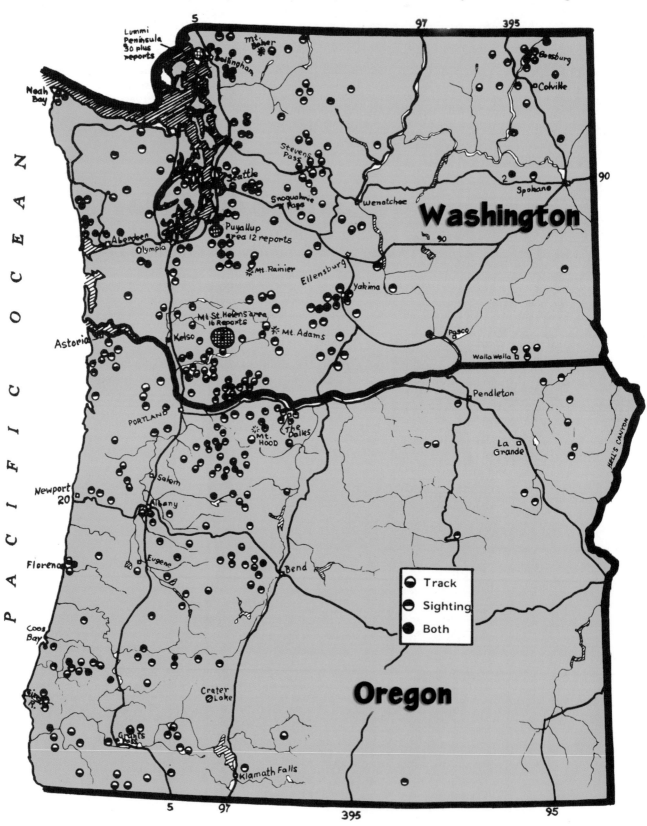

California
Sighting and track reports in Northern California

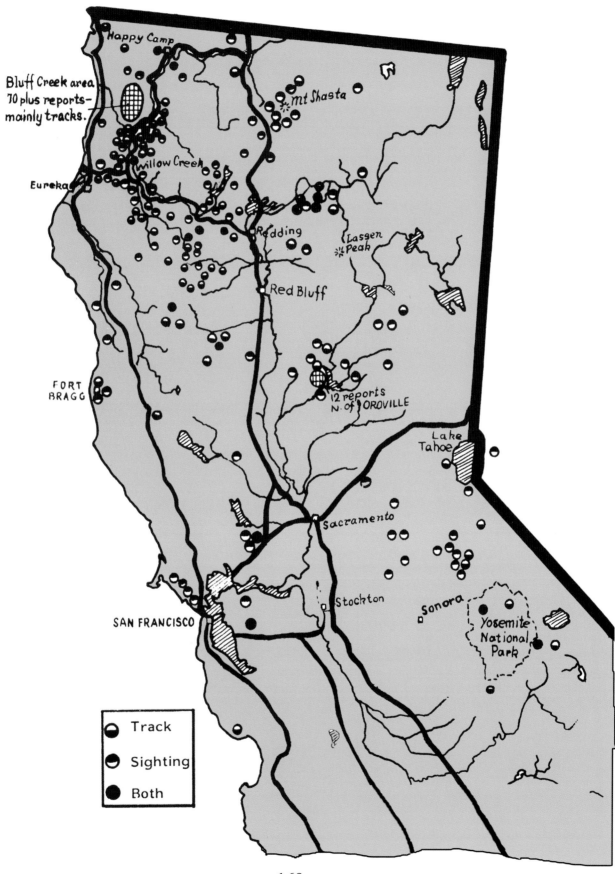

Montana and Idaho
Sighting and track reports in Montana and Idaho

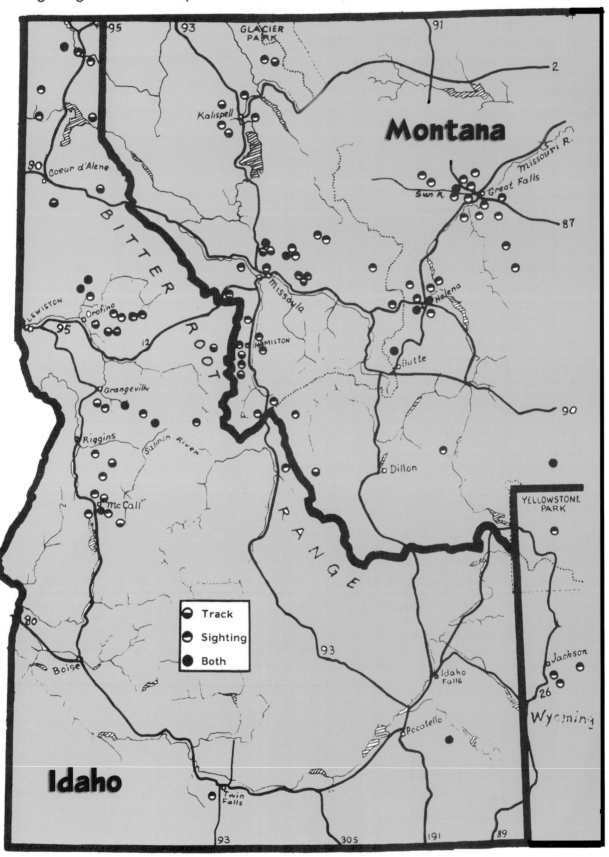

The following map and adjacent chart shows the number of sasquatch related reports throughout North America as of 2003. I compiled the figures using statistics available from John Green, the Bigfoot Field Researchers Organization (BFRO) and Yvon Leclerc. What I show are simple reported incidents "on record." I have not used any other standard for inclusion of an incident. Figures arrived at by John Green and the BFRO individually will differ because their standards differ. It must be kept in mind that the figures shown are *reported* incidents. Certainly many incidents are not reported so the figures reflect the minimum number of incidents. The total number of reports shown on this map is 2,557. The time span is effectively 100 years.

N. A. REPORTS

P/S	NBR.
AB	70
BC	362
MB	35
NB	5
NL	1
NT	2
NS	0
NU	0
ON	25
PE	0
QC	26
SK	3
YT	2
TOTAL	**531**
AL	23
AK	20
AZ	16
AR	32
CA	343
CO	60
CT	3
DE	1
FL	104
GA	20
HI	0
ID	32
IL	23
IN	30
IA	21
KS	16
KY	31
LA	24
ME	11
MD	18
MA	5
MI	49
MN	21
MS	9
MO	26
MT	74
NE	6
NV	5
NH	5
NJ	36
NM	12
NY	53
NC	20
ND	2
OH	95
OK	33
OR	176
PA	58
RI	2
SC	20
SD	9
TN	29
TX	63
UT	27
VT	4
VA	14
WA	286
WV	18
WI	20
WY	21
TOTAL	**2026**
G.TOT.	**2557**

Sasquatch Related Reports in North America as of 2003

SASQUATCH AND THE SMITHSONIAN INSTITUTION

With all of the evidence we have indicating the reality of sasquatch, one would think that the Smithsonian Institution would be "hot on the trail." Sadly, such is not the case, despite the reasonably positive opinions of Dr. John Napier, head of the Smithsonian primate program in the early 1970s.

The Patterson/Gimlin film was shown to the Smithsonian anthropologists and other scientific people in 1969. To my knowledge, there was no official report provided on the opinions and feelings of the people who attended the screening. However, news of what went on "behind closed doors" was communicated by at least one of the attendees and found its way to sasquatch researchers. I have no way of determining the validity of the information, but judging by the comments made by scientists at the University of British Columbia screening two years earlier, it appears plausible.

> ". . . what went on 'behind closed doors' was communicated by at least one of the attendees and found its way to sasquatch researchers."

The following is a summary of what is believed were the opinions/thoughts of the Smithsonian people. We are told about twenty of their professionals attended the screening and that they were shown "movies and stills." By the plural word "movies" it appears both film rolls taken by Patterson were screened.

1. The creature was supposed to be a female, yet they all agreed that it walked not merely like a human being, but like a man, a male human being.

2. The creature had a sagittal crest (pointed head) which is not merely an ape characteristic, but a feature of a male ape, and only a mature adult male at that. They reasoned that an uninformed hoaxer would probably make this mistake because gorilla costumes are usually that of male gorillas. Female costumes would not be provided unless specifically requested. *The inference here being that the hoaxer used a male gorilla costume (i.e., a gorilla is a gorilla) and changed it to a female by adding breasts. Being uninformed, the hoaxer failed to realize that females do not have a sagittal crest. They went on to reason that a large man inside such a suit, padded out in the shoulders and bust, would be shaped like a giant female, but would give himself away (at least to experts) by his male gait, even without the male sagittal crest.*

3. One physical anthropologist who specializes in bone structure, stated that judging from the footprint (note singular term) left by the creature, the toes were too short for the length of the foot. *(I believe the reference her is to a cast of the footprint which would have likely been provided. There are no clear footprints seen in the film showing the creature. However such are seen in the second film roll and it is possible stills (ordinary photographs) were provided of actual footprints (i.e., the Laverty photographs).. Nevertheless, if either were the case I think the plural term "footprints" would have been used).*

4. A comment was made on the odor Patterson reported, quoting from the source: "What creature could 'smell terrible' to a human nose at a distance of a hundred feet in open air?"

5. Some thought was apparently given to the cost of making a costume for the purpose of a hoax. It appears an estimate of such a cost was provided. The comment (or consensus) was that while an ape man costume would be expensive, it would be nothing like the price quoted. The opinion was that the price quoted was probably for the provision of a mechanical sasquatch.

6. One scientist "allowed the possibility of the film being genuine" (i.e., shows a natural creature) even though he mentioned reservations.

The information provided raises four (4) specific issues concerning the nature of the creature:

Point 1: The unusual walk
Point 2: The sagittal crest
Point 3: The short toes
Point 4. The odor

The first two points were raised at the University of British Columbia screening and have been addressed by John Green in the previous section, **The First Film Screening to Scientists**. John has summarized findings on these concerns. *They definitely do not detract from the credibility of the creature filmed.*

The third point (short toes) has been addressed by Dr. Jeffrey Meldrum. His conclusion on the issue is quoted here: *The Bluff Creek tracks (film site tracks) don't really have short toes. They simply appear short at first glance due to a slightly more extensive sole pad at the base of the toes. Closer examination reveals the presence of a flexion crease that marks the position of the hallucial-metatarsal joint at the base of the big toe, which position is consistent with tracks from elsewhere that appear to have longer toes. This also explains the apparent "double ball" feature that is present in a few of the Bluff Creek tracks but not evident elsewhere.*

The fourth point (odor) has not been addressed by any professionals or others to my knowledge. I do not know of research

A Little Peep Outside the Closet

In January 1974, the Smithsonian people did emerge from their closet and take a quick look around. In that month, an article entitled, The Search Goes on for Bigfoot, appeared in the institution's magazine, *The Smithsonian*.

The article provides a general summary of the then current status in the Bigfoot arena - nothing new, profound or even promising. The name of the author is not shown, however, he or she did make a marginal admission. In discussing the bigfoot issue with a nature editor in New York, the author was told that a sasquatch could not possibly remain hidden in this day and age when the woods are crawling with hunters, campers and snowmobilers. The author then states:

"Shortly thereafter I was flying down through the river valleys of northern California and the thought crossed my mind that you could hide a herd of elephants in any square mile of that country with no trouble at all."

If this author were to fly over Canada's wilderness and remote areas, he or she would see that we would have no trouble at all hiding the entire State of California!

Dr. John Napier: "I am convinced that the Sasquatch exists, but whether it is all that it is cracked up to be is another matter altogether."

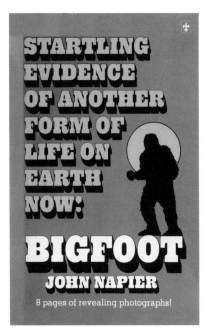

performed in this regard. All I can say is that some other sighting reports state there was an odor. It should be noted that when the creature in the film was first spotted, according to Bob Gimlin it was only about 50 feet/15.2m away.

As to the Smithsonian thoughts on the cost of a costume, I do not know what estimate they were given. It is, however, amusing that the prospect of a "mechanical sasquatch" is even mentioned. We don't even have the technology at this time to produce a mechanical sasquatch like that seen in the film. The point made is totally absurd. I really can't believe that any Smithsonian scientist would entertain this thought.

The scientist who allowed the possibility of the film being genuine is undoubtedly Dr. John Napier (seen here) who went so far as to write a book on bigfoot. The book, entitled *Bigfoot: Startling Evidence of Another Form of Life on Earth Now,* was published in 1972 (Berkley Publishing Corporation). Napier was at one time Director of the Primate Biology program at the Smithsonian Institution. At the time he wrote his book (1972) he was visiting professor of Primate Biology at the University of London. He was a British physical anthropologist with a dry wit and excellent writing ability, so the book is highly informative and entertaining at the same time. However, it is only on the last page of the book that he actually professes a "qualified" belief in the creature. The following are his words:

I am convinced that the Sasquatch exists, but whether it is all that it is cracked up to be is another matter altogether. There must be something in north-west America that needs explaining, and that something leaves manlike footprints. The evidence I have adduced in favor of the reality of Sasquatch is not hard evidence; few physicists, biologists or chemists would accept it, but nevertheless it is evidence and cannot be ignored.

Nevertheless, in the preface to the 1976 edition of this book (now titled *Bigfoot, the Yeti and Sasquatch in Myth and Reality*, Napier states:

One is forced to conclude that a manlike life-form of gigantic proportions is living at the present time in the wild areas of the northwestern United States and British Columbia. If I have given the impression that this conclusion is - to me - profoundly disturbing, then I have made my point. That such a creature should be alive and kicking in our midst, unrecognized and unclassifiable, is a profound blow to the credibility of modern anthropology.

Napier is no longer with us and the Smithsonian has remained distant on the issue to this day.

SASQUATCH PROTECTION

On April 1, 1969 the Board of Commissioners of Skamania County, Washington State, adopted an ordinance for the protection of sasquatch/bigfoot creatures. The choice of the adoption day (April Fools Day) may have been intentional, reflecting a little "light heartedness." Nevertheless, it was an official ordinance and is still in effect. *However, it has been partially repealed and amended because it "may have" exceeded the jurisdictional authority of the Board of Commissioners.* The revised ordinance went into effect on April 2, 1984. The full text of the original ordinance and its aftermath is as follows. I then present the full text of the revised ordinance.

ORDINANCE NO. 69-01

Be it hereby ordained by the Board of County Commissioners of Skamania County:

WHEREAS, there is evidence to indicate the possible existence in Skamania County of a nocturnal primate mammal variously described as an ape-like creature of a sub-species of Homo Sapiens, and

WHEREAS, both legend and purported recent sightings and spoor support this possibility, and

WHEREAS, this creature is generally and commonly known as a "Sasquatch," "Yeti," "Bigfoot," or Giant Hairy Ape," and

WHEREAS, publicity attendant upon such real or imagined sightings has resulted in an influx of scientific investigators as well as casual hunters, many armed with lethal weapons, and

WHEREAS, the absence of specific laws covering the taking of specimens encourages laxity in the use of fire arms and other deadly devices and poses a clear and present threat to the safety and well-being of persons living or traveling within the boundaries of Skamania County as well as to the creatures themselves,

THEREFORE BE IT RESOLVED that any premeditated, willful and wanton slaying of any such creature shall be deemed a felony punishable by a fine not to exceed Ten Thousand Dollars ($10,000) and/or imprisonment in the county jail for a period not to exceed Five (5) years.

BE IT FURTHER RESOLVED that the situation existing constitutes an emergency and as such this ordinance is effective immediately.

ADOPTED this 1st day of April, 1969.
Board of Commissioners of Skamania County. By: CONRAD LUNDY JR, Chairman.
Approved: ROBERT K. LEICK, Skamania County Prosecuting Attorney.
Publ. April 4, 11, 1969

The Story Behind the Skamania County Ordinance

This ordinance was brought about as a direct result of a complaint by the noted sasquatch researcher, Robert W. Morgan. As it happened, Morgan, with his friends Leonard Cairo and Bill Tero, went to Stevenson (Skamania County) to follow-up on a sasquatch sighting. Sheriff Closner's deputies had made some casts of footprints associated with the incident so there had been considerable publicity on the sighting.

When Morgan arrived in the town he found it crowded with hunters who were literally blocking some traffic. Morgan met with Conrad Lundy, the Skamania County Commissioner and said that something should be done about the situation. Lundy said such was not against the law and there was little he could do. Morgan angrily insisted that there should be a law and he should get one passed before summer set in because "those damned fools were going to shoot something they knew nothing about, and to make matters worse, they made it dangerous for ordinary people." Lundy stared at Morgan in silence for a short time, evidently reflecting on the concerns he expressed. Lundy than asked Morgan to go with him to meet with Roy Craft of the *Skamania County Pioneer* newspaper. The little group huddled for nearly an hour and drafted the outline for what became the now famous Skamania County Ordinance No. 69-01.

Sheriff Bill Closner holds a 22-inch/55.9cm cast he made of a Skamania County footprint.

Sheriff Bill Closner (left) and Deputy Jack Wright study a print found in Skamania County, Washington, March 1969. The two officers were responding to a report by a couple who found large footprints on their recreational property adjacent to Bear Creek. A few days earlier, the sheriff had received a report of a sasquatch sighting by Don Cox of Washougal. Cox stated that in the early morning of March 5, 1969 he seen a haircovered upright creature run across the highway near Beacon Rock State Park. He said the creature was between 8 and 10 feet (2.4 and 3.1 m) tall and ran like a man but was covered with fuzzy fur and had the face of an ape. While there is no distinct connection between the two events, such appears possible.

The ordinance was published in the local weekly newspaper, *Skamania County Pioneer*, on April 4 and April 11, 1969. It appears, however, that people were not convinced that the ordinance was serious, being undoubtedly influenced by the adoption date of April 1. Consequently, the newspaper publisher had the article notarized on April 12, 1969 and printed both the ordinance and an Affidavit of Publication in a subsequent paper edition under the heading, **Here's Notarized Text of Skamania County's Bigfoot Ordinance**. The affidavit is shown below.

Revised Ordinance

ORDINANCE NO. 1984-2
PARTIALLY REPEALING AND AMENDING ORDINACE NO. 1969-01

WHEREAS, evidence continues to accumulate indicating the possible existence within Skamania County of a nocturnal primate mammal variously described as an ape-like creature or a sub-species of Homo Sapiens; and

WHEREAS, legend, purported recent findings, and spoor support this possibility; and

WHEREAS, this creature is generally and commonly known as "Sasquatch", "Yeti", "Bigfoot", or "Giant Hairy Ape", all of which terms may hereinafter be used interchangeably; and

WHEREAS, publicity attendant upon such real or imagined findings and other evidence have resulted in an influx of scientific investigation as well as casual hunters, most of which are armed with lethal weapons; and

WHEREAS, the absence of specific national and state laws restricting the taking of specimens has created a dangerous state of affairs within the county with regard to firearms and other deadly devices used to hunt the Yeti and poses a clear and present danger to the safety and well-being of persons living or traveling within the boundaries of this country as well as the Giant Hairy Apes themselves; and

WHEREAS, previous County Ordinance No. 1969-01 deemed the slaying of such a creature to be a felony (punishable by 5 years in prison) and may have exceeded the jurisdictional authority of that Board of County Commissioners; now, therefore

BE IT HEREBY ORDAINED BY THE BOARD OF COUNTY COMMISSIONERS OF SKAMANIA COUNTY THAT THAT PORTION OF Ordinance No. 1960-01, deeming the slaying of Bigfoot to be a felony and pun-ishable by 5 years in prison, is hereby repealed and in its stead the following sections are enacted:

Section1. Sasquatch Refuge. The Sasquatch, Yeti, Bigfoot, or Giant Hairy Ape are declared to be endan-gered species of Skamania County and there is hereby created a Sasquatch Refuge, the boundaries of which shall be co-extensive with the boundaries of Skamania County.

Section 2. Crime - Penalty. From and after the passage of this ordinance the premeditated, willful, or wanton slaying of Sasquatch shall be unlawful and shall be punishable as follows:

(a) If the actor is found to be guilty of such a crime with malice aforethought, such act shall be deemed a Gross Misdemeanor.

(b) If the act is found to be premeditated and willful or wanton but without malice aforethought, such act shall be deemed a Misdemeanor.

(c) A gross misdemeanor slaying of Sasquatch shall be punishable by 1 year in the county jail and a $1,000 fine, or both.

(d) The slaying of Sasquatch which is deemed a misdemeanor shall be punishable by a $500.00 fine and up to 6 months in the county jail, or both.

SECTION 3. Defense. In the prosecution and trial of any accused Sasquatch killer the fact that the actor is suffering from insane delusions, diminished capacity, or that the act was the product of a diseased minde, shall not be a defense.

SECTION 4. Humanoid/Anthropoid. Should the Skamania County Coroner determine any victim/creature to have been humanoid the Prosecuting Attorney shall pursue the case under existing laws pertaining to homicide. Should the coroner determine the victim to have been an anthropoid (ape-like creature) the Prosecuting Attorney shall proceed under the terms of this ordinance.

BE IT FURTHER ORDAINED that the situation existing constitutes an emergency and as such this ordi-nance shall become effective immediately upon its' passage.

REVIEWED this 2nd day of April, 1984, and set for public hearing on the 16th day of April, 1984, at 10:30 o'clock a.m.

BOARD OF COUNTY COMMISSIONERS
Skamania County, Washington

(Signed by the Chairman, two Commissioners and the County Auditor and Ex-Officio Clerk of the Board)

We also have a resolution adopted by Whatcom County, Washington that declares the county a sasquatch protection and refuge area. The resolution went into effect in June 1992 (Resolution No. 92-043).

SASQUATCH ROOTS

If sasquatch do indeed exist, the main question to be answered is, what kind of creature is it? Certainly, the only way this question can be properly answered is by having an actual body of the creature (or body part) or at the very least, bones. Despite a few alleged killings of sasquatch and at least one report of a rotting carcass, we still don't have any evidence of this nature. Given the evidence we do have, considerable speculation has been made on the creature's true identity. The most popular theory is that sasquatch belong to a species called *Gigantopithecus blacki* that was assumed to have become extinct about 300,000 years ago. Evidence of this creature's existence is based on jawbones and teeth found in China and India. *Gigantopithecus blacki* is the largest primate that has ever been known to exist and as such becomes a reasonable candidate for sasquatch. It is speculated that some of the ancient creatures found their way over the land bridge that once connected Eurasia with North America (previously discussed). In their new domain, these prehistoric immigrants apparently flourished and were not affected by the conditions that caused the extinction of their relatives who remained in Eurasia. Dr. Grover S. Krantz was the main proponent of the *Gigantopithecus blacki* theory. He is seen in the opening photograph with a model of the creature. Based on a lower jawbone, Dr. Krantz constructed the entire skull of a *Gigantopithecus blacki* which is seen in the next photograph compared with a gorilla skull and human skull.

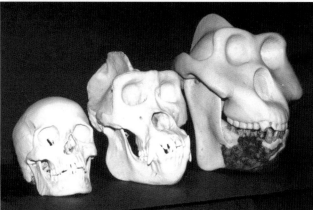

HUMAN   GORILLA   GIGANTO B

The following photographs and captions were reprinted from *Bigfoot/Sasquatch Evidence,* by Dr. Grover Krantz, (Hancock House Publishers, 1999).

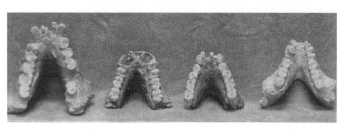

Four Gigantopithecus jaws. Starting from the left, these are the adult male, young male and adult female - all from China. At the far right is the adult female from India. The Asian ape was probably bipedal, and in every known and surmised characteristic is an exact match for sasquatch.

Gigantopithecus size contrasts. The adult male from China is conspicuously larger than a large male orangutan (center) which is almost the size of a male gorilla jaw. The corresponding part of a big man's jaw (left) is tiny by comparison.

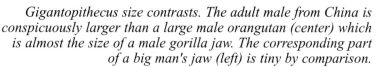

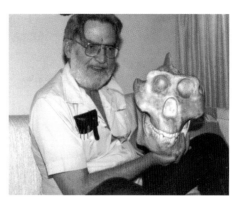

John Green (left) is seen here with Dan Murphy holding a copy of the Gigantopithecus blacki skull created by Dr. Grover Krantz. (Photo: 1994)

Dr. Grover Krantz with his skull model

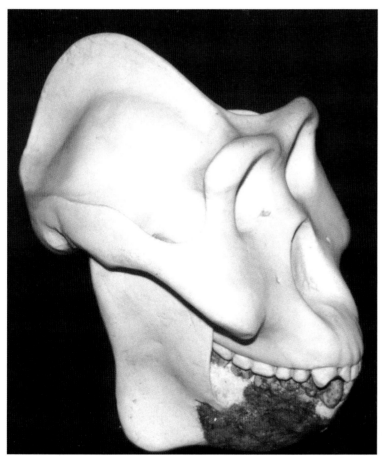
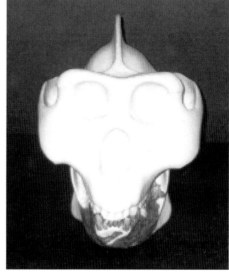
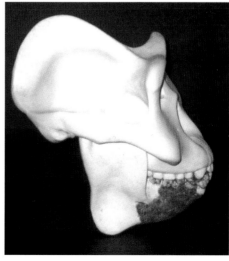

This replica of Dr. Krantz's model was made by BoneClones, California. It is available for purchase and is a highly prized item in the field of sasquatch research.

Tributes - American and Canadian Researchers

During the last ten years, I met and associated with many highly dedicated American and Canadian sasquatch researchers. There are indeed many, many more with whom I have interacted on a casual or limited basis . . .

Strictly from my own experience, I have prepared the following tributes to those researchers who have established and sustained a continuing presence in the field of sasquatch studies. These people pioneered the research, wrote books, published newsletters, established organizations/societies, and created major internet web site. I deemed it both appropriate and essential that the reader have some first-hand knowledge of these individuals and the work they have done.

It needs to be stressed that what I present here is not all-inclusive. Scientists such as Dr. Henner Fahrenbach, Dr. D. Jeffrey Meldrum, Dr. John Bindernagel and my Russian hominologist associates, Dmitri Bayanov and Igor Bourtsev (who have contributed volumes to North American sasquatch studies) are featured in their own section or sections connected with their important findings. The same holds true for Bob Gimlin, Ron Morehead, Robert Morgan and my Ohio friends, Joedy Cook and George Clappison.

I recently acquired the work by Thom Powell, *The Locals: A contemporary Investigation of the Bigfoot/Sasquatch Phenomenon* (Hancock House Publishers, 2003), which is certainly destined to become another classic in the annals of sasquatch studies. I have only "crossed paths" with Thom (conference in Oregon) and look forward to seeing him again.

There are other researchers beyond my scope of interaction who have also made very significant contributions but I can only speak from my personal knowledge thereby severely limiting the range of tribute inclusions.

Those other researchers and notables with whom I have had some personal contact and deserve mention are as follows: Janet Bord, Constance Cameron, Jimmy Chilcutt, Loren Coleman, M.K. Davis, George Early, Henry Franzoni, Craig Heinselman, Jo Ann Hereford, Don Keating, Larry Lund, Jim McClarin, Lee McFarland, Bob Montgomery, Sally Newberry, Todd Neiss, Michael Quast, Dr. Peter Rubec, Michael Rugg and Ron Schaffner.

As we move ahead and more people take up the cause to resolve the sasquatch issue, I look forward to future associations. I certainly feel we are on the brink of a major discovery and encourage all researchers to continue their valuable work.

THE INCOMPARABLE BOB TITMUS

Bob Titmus, the greatest of the 20th century "sasquatch hunters" is generally thought of as an American, but he spent two thirds of his adult life in British Columbia - on the northern coast, in the Hazelton area and at Harrison Hot Springs. He died a Canadian citizen. Not well-known to the public, because he never sought publicity or wrote a book, he devoted more time to actually hunting for sasquatch creatures and had more to show for his efforts, than anyone else.

He was a taxidermist at Anderson, California, when huge humanlike tracks started showing up on a road under construction in the Bluff Creek valley during the summer of 1958. He showed his old friend Jerry Crew how to make the plaster cast that introduced "Bigfoot" to the world. That cast, and subsequent examination of the tracks themselves, convinced Bob that a real creature had to be out there; something he spent the rest of his life trying to prove.

Bob's greatest success came within a few weeks of the Jerry Crew incident. Bob and a friend found slightly smaller tracks of a distinctly different shape on a sandbar beside Bluff Creek, proving that a species of animal was involved, not a freak individual. Bob made casts of these tracks and subsequently found those same tracks again at more than one location. He made more casts of the familiar tracks and went on to cast other tracks he later found in California, Oregon and on islands off the central coast of British Columbia (B.C.). All casts he made are among the best ever made anywhere. Bob was also called in to examine and cast tracks found by other people. Although most of his B.C. material was lost when his boat burned, his collection was by far the largest collection of original casts made by any individual.

In 1959, Bob, along with John Green and René Dahinden, persuaded Texas millionaire Tom Slick to finance a full-time bigfoot hunt in northern California. However, the hunt produced only some more footprints so in 1961 Bob shifted his efforts to what seemed at the time to be a more promising area – that centering on Klemtu, B.C. This venture also petered out, but Bob found life in northern B.C. to his liking and he stayed on, settling at Kitimat and

Bob was a highly meticulous and methodical man. Everything he did from letter-writing to making plaster casts was perfect.

HISTORICAL SIDE NOTE:
It was Bob Titmus who sparked an interest in sasquatch with biologist David Hancock. As it happened, Hancock was studying bald eagles and the white spirit bears along the central B.C. coast. He met Titmus on one of his outings and helped him check his network of trip-wired cameras. Hancock was highly impressed with Titmus' sincerity and passion for the sasquatch. When Hancock later became a publisher he elected to publish books related to the sasquatch and has become the major publisher of such books – which, of course, includes this book.

later Hazelton, B.C. In 1977 he investigated tracks found by children near the Skeena River and made the best set (left and right foot) of casts ever obtained.

Bob told of two personal sasquatch sightings on the northern coast, one during World War II from a ship in Alaskan waters (he said he had refused to credit his senses at the time); and one of three dark bipeds scaling a cliff a long way off near Kitimat. This sighting occurred while he was searching with his own boat in the 1960s. Like most witnesses he had no proof of these experiences. Nevertheless, he did find and cast footprints corroborating reports by others, including a remarkable set of casts of the prints left by the creature in the Patterson/Gimlin film. One of these casts showed that the creature's foot could bend in the middle in a way not possible for a human foot.

By the time Bob moved to Harrison Hot Springs in 1978, his field work was restricted by health problems, including increasing pain from a back injury he had suffered keeping his boat from going on the rocks in a storm. However, he continued to investigate reports in the nearby Fraser Valley and also to hunt for more evidence at Bluff Creek. On one of his trips there, he collected brown hairs from branches where there was evidence that a sasquatch had passed. These hairs were later proven to be from a higher primate but defied specific identification. On another trip he drained a large pond in order to make a cast of a sasquatch hand print.

Bob Titmus died at Chilliwack, B.C. in 1997 and his ashes were scattered on a Harrison Lake mountainside. His American material is now displayed in a wing of the Willow Creek – China Flat Museum in California that was built especially to house it.

Bob taking a break. I missed the opportunity to meet him, something I very much regret to this day. This photograph was taken on Bob's last trip to Bluff Creek in the 1990s.

Bob is seen here (left) with Dr. Grover Krantz, sometime in the 1990s.

The Willow Creek – China Flat Museum is seen here in a photograph taken in September 2003.

JOHN GREEN - THE LEGEND AMONG US

Without doubt, John Green is the pre-eminent authority on the sasquatch/bigfoot issue. Although no longer very active in field research, he spent many years investigating sasquatch sightings and footprint reports. He has hunted for evidence in many remote areas throughout the Pacific Northwest and has traveled to eastern Canada and throughout the United States, methodically documenting and photographing evidence of the creature's existence. He has personally interviewed hundreds of people including all of the early sasquatch witnesses - Albert Ostman, Fred Beck, Roger Patterson, Bob Gimlin, to name a few. John has authored several books on the subject, the most noteworthy being his *Sasquatch, The Apes Among Us*. In the "sasquatch fraternity," as it were, John Green is the "clearing house" for all matters. His vast knowledge in the field has no equal.

John became involved in investigating the sasquatch in 1957 while he was owner/publisher of the *Aggassiz-Harrison Advance* newspaper. The Harrison area was noted for sasquatch sightings. John, however, took little interest in the subject until he learned that some people in the community he had come to respect had been witnesses to an incident at nearby Ruby Creek 16-years earlier. Thereupon, John teamed up with René Dahinden, who had come to Harrison to hunt for the sasquatch, and the two men embarked on serious and dedicated research. Both John and René were founding members of the Pacific Northwest Expedition in California in 1959. John and René continued to cooperate with each other for over a decade. Eventually they parted company over the issue of sharing information with other people. René never wavered from his determination to solve the mystery himself. John gave up on that prospect and does whatever he can to help anyone he considers to be making sincere efforts in sasquatch research.

John has documented thousands of sasquatch related incidents, many of which he personally investigated. He has diligently analyzed the information he has collected and has provided many statistics on the nature and distribution of the creature. He has presented his findings at numerous conferences and continues to be called upon for speaking engagements. Mainly through John's efforts, some highly eminent anthropologists and zoologists are now involved in the sasquatch issue.

I met John in 1993. Over the years I have visited him a number of times and have listened intently to his views. Few things escape his notice. He is exceedingly critical and accepts absolutely nothing at face value. John is very careful where doubt is involved

John Green is seen here in about 1973. By this time he had been involved in sasquatch research for some 14 years and was both well-known and highly respected in the field.

"John has documented thousands of sasquatch related incidents."

John Green has documented more information on the saquatch than anyone. His books never fail to inspire readers – believers and non-believers alike.

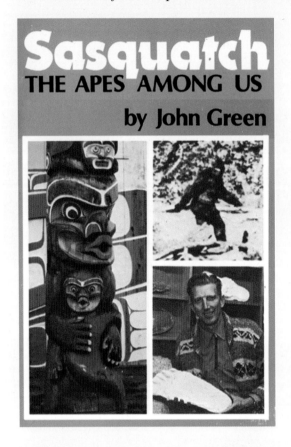

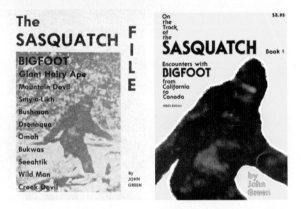

John Green's books (not all titles are shown) are the cornerstone of sasquatch research. Over 200,000 copies have been sold. His latest book (last shown) combines two previous books and provides extensive review of the controversy surrounding the infamous hoaxer, Ray Wallace. It also provides the latest scientific developments in the field of sasquatch studies..

with people and sasquatch related evidence. Indeed, John gives little or no credibility to most of the highly publicized sasquatch encounters (videos in particular) of recent years. He is highly uncomfortable with all findings by Ivan Marx and Paul Freeman. It is only through my insistence that some material provided by these individuals is included in this work.

John is seen on the right in this photograph taken in April 2003 with Lynn Maranda and your author. Lynn is curator of anthropology for the Vancouver Museum. She and Lee Drever (inset), the museum's marketing and communications administrator, accompanied me on a visit to see John to discuss the sasquatch exhibit. John was highly supportive from the outset on the prospects of having an exhibit and was the foremost contributor of exhibit artifacts, knowledge and photographs for this work.

This very early photograph of John was taken when he was about 30 years old. John was a graduate of the University of British Columbia and the Columbia University Graduate School of Journalism in New York. He worked with daily newspapers in Toronto, Victoria and Vancouver before buying the weekly Agassiz-Harrison Advance newspaper.

In California (1967) with White Lady, a tracking dog.

John is seen here at his newspaper office in 1970 with stacks of questionnaire booklets used to gather information on sasquatch related incidents for a computer-based study.

In October 1997, John attended a conference on hominology in Moscow, Russia. He is seen here on the right with Dr. Grover Krantz (left) and Igor Bourtsev at Igor's apartment. Igor is holding a statue he created of the creature seen in frame 352 of the Patterson/Gimlin film.

John with George Haas in Oakland, California in the spring, 1976. George was publisher of the Bigfoot Bulletin, the first newsletter in the sasquatch field of study.

Left to right: Dr. Jeff Meldrum, Tom Steenburg, Dr. John Bindernagel and John Green - at a Harrison Hot Springs Sasquatch Symposium.

John Green in his office, July 2003.

John is seen here on the right as a member of the Pacific Northwest Expedition in 1959. The photograph was taken at Louse Camp, Bluff Creek and Notice Creek area, California. Tom Slick, the expedition financer took the photo. The other members shown from left to right are: Bob Titmus, Ed Patrick, Geri Walsh, René Dahinden and Kirk Johnson.

ON THE ROAD TO WILLOW CREEK. *(L to R) Bob Gimlin, John Green, Chris Murphy and Dmitri Bayanov enroute to the Willow Creek, California Bigfoot Symposium, September 2003. Tom Steenburg (inset), who also traveled with the group, took the photograph. I was able to spend a lot of valuable time with Bob and Dmitri on this trip. This was the first time I had met them in person, although I have corresponded with Dmitri in Russia for many years.*

THE INTREPID DR. GROVER S. KRANTZ

Dr. Grover S. Krantz (d. 2002) was a physical anthropologist with Washington State University. He became involved in the sasquatch issue in 1963 and spent the next thirty-nine years relentlessly investigating the evidence provided to him. He found what he considered indisputable evidence in dermal ridges (i.e., like finger prints) that he discovered on some footprint casts. In his own words: *"When I first realized the potential significance of dermal ridges showing in sasquatch footprints, it seemed to me that scientific acceptance of the existence of the species might be achieved without having to bring in a specimen of the animal itself. It was this hope that drove me to expend so much of my resources on it, and of my scientific reputations as well."* Unfortunately, Dr. Krantz was not able to bring about the scientific acceptance he envisioned. Nevertheless, it was confirmed by Jimmy Chilcutt, a finger print expert who has made a special study of the dermatoglyphics on the hands and feet of nonhuman primates, that dermal ridges discovered by Krantz indicate they are definitely those of a nonhuman primate.

Dr. Krantz's most notable sasquatch related accomplishment was his reconstruction of the skull of *Gigantopithecus blacki*, an extinct primate that lived in southern China somewhere between 500,000 and 1,000,000 years ago. Krantz theorized that the sasquatch may have descended from this primate. His model is based on a lower jaw fossil of the creature. A full discussion on this aspect is provided under **SASQUATCH ROOTS.**

Dr. Krantz was also the major supporter for the authenticity of the Bossburg "cripple foot" casts. He studied these casts intently and provided a proposed bone structure for each cast. Despite skepticism, these casts are very intriguing.

These books written by Dr. Krantz present his findings in full detail. The first book was published in 1992. The second book, an update, was released in 1999 by Hancock House Publishers.

Although highly regarded as an anthro-pologist, Dr. Krantz's reputation was greatly diminished because of his belief in the reality of the sasquatch. While certainly bothered by this eventuality, he did not let it stop him in any way. He spoke at sasquatch symposiums, appeared in many television documentaries and was continually quoted in newspaper articles. I had the pleasure of meeting Dr. Krantz on two occasions. He was a very gentle man, very approachable and friendly. While he was supportive of action to intentionally kill a sasquatch to definitely establish the creature's existence, I really wonder if he personally would have been able to "pull the trigger."

As with Roger Patterson, I think the day will come when Dr. Krantz will be officially and properly recognized by the scientific community for his research on the sasquatch.

Here we see Dr. Krantz in his laboratory. Despite the mountain of evidence he uncovered and published few scientists were willing to even consider his findings. Most of his evidence and specimens now reside at the Smithsonian Institute, Washington, D.C.

In 1997, Dr. Krantz attended a conference on hominology in Moscow, Russia. He is seen here at a social gathering. From left to right: Vadim Makarov, Dmitri Bayanov, Igor Bourtsev, Marie-Jeanne Koffmann, Grover Krantz, Dmitri Donskoy, Michail Prachtengertz and in the foreground, Alexandra Bourtseva. John Green took the photograph.

Dr. John Bindernagel, left, and Dr. Krantz, 1999.

Dr. Krantz speaking at the 3rd International Sasquatch Symposium held in Vancouver, B.C., September 1999. The symposiums were organized by Stephen Harvey. Dr. Krantz was always a keynote speaker at symposiums and indeed the center of attention. He never failed to shed new light on the sasquatch issue.

THE RELENTLESS RENÉ DAHINDEN

René Dahinden was so closely associated and involved in the sasquatch phenomenon, and so widely publicized in this connection, that his very name brings the creature to mind. Dahinden was born in Switzerland in 1930 and came to Canada in 1953. About two months after his arrival he and his employer, Wilbur Willich, listened to a CBC Radio program about a *Daily Mail* (British newspaper) expedition to search for the yeti. René remarked, "Wouldn't it be something to be part of that!" Willich replied, "You don't need to go that far, we have the same things right here." Dahinden then learned that Canada (British Columbia specifically) had its own version of the yeti - our elusive sasquatch. From that point on René became obsessed with finding the creature. He headed for British Columbia and later spent months at a time wandering through the province's vast wilderness in search of his prey. In time he became well known as a "sasquatch hunter." When not in the bush, he responded to sasquatch sighting reports all over the Pacific northwest. He interviewed hundreds of people and amassed a formidable collection of sasquatch related artifacts and literature.

I became associated with René in 1993, about forty years into his search. He was now 63-years old and was no longer spending much time in the field. He lived just a few miles from where I live and over the next five years or so, I visited Rene two or three times a week. I spent many long evenings with him discussing the sasquatch issue. Although René was a firm believer in the existence of the creature, he never saw one. Nevertheless, he was a highly dedicated and diligent researcher right to the end.

René is shown here in a photograph I took in 1994. Below is shown "René's book"- the first (1973), second (1975), and third (1993) editions (book written by Don Hunter with René). While frames from the Patterson/Gimlin film adorn the covers, few film frames were used in the book. When I questioned René as to why he did not use more film material, he told me that the book was about the sasquatch in general. In other words the book is fully balanced on the issue - excessive film material would detract from other highly credible, but less illustrative evidence.

René measuring prints on Blue Creek Mountain, California, August 1967. At that time René and John Green were close friends and worked together on the Blue Creek Mountain findings.

René, left, is seen here with Roger Patterson and his Welsh ponies in 1967. Patterson could transport two ponies in his Volkswagen van. The sign on top of the van advertises Roger's "prop lock" invention - a locking device used on fruit tree props.

René, right, discussing plans with Clayton Mack, the famous grizzly bear guide, at Anahim Lake (B.C.) Stampede (1960s or 70s).

René, left, with Roger Patterson at Patterson's home, Tampico, Washington, in the spring of 1967. In the fall of that year, Patterson took his famous movie footage of a sasquatch creature at Bluff Creek, California.

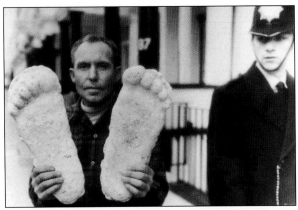

René in London, England in November 1971. He also visited Finland, Sweden, Switzerland and Russia to draw European scientific attention to the sasquatch issue.

René, left, and Ivan Marx at Ivan's home, Burney, California, spring 1967.

René's two sons, Martin (left) age 9 and Erik age 15, are seen here with Jim McClarin's sasquatch carving, now located on the grounds of the Willow Creek Museum. (Photo taken in the early 1970s.)

René gazes out over the mountainous countryside in B.C.'s Garibaldi Park.

René, left, with Ivan Sanderson in New York, 1971.

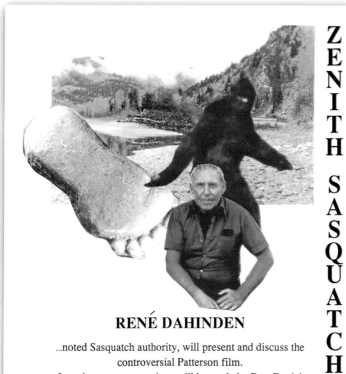

Rene cooling his heels in Harrison Lake, summer 1995.

In early 1995, René agreed to put on a presentation for my lodge. April 20, 1995 was set and I posted this ad throughout the greater Vancouver Masonic community. The response was remarkable–one of the best attended events in the lodge's history. René showed the Paterson/Gimlin film (the real thing) and many slides.

René in a particularly jovial mood. He really liked to joke around with Dan.

Dan Murphy, René and Chris Murphy Jr. are seen in this photographs at René's place in 1994. Dan and I later worked with René in making posters from the Patterson/Gimlin film frames and marketing footprint casts.

René had his "Land of the Sasquatch" printed in 1986. The idea for the sequence appears to have been taken from an article entitled, "The Trail of the Sasquatch," featured in **The Standard–Weekend Magazine** (Montreal), April 11, 1959. Jack Davis, the artist who illustrated the article, also provided the artwork for René's pamphlet. The story line for the magazine article has an unusual history and was part of a significant chain of events. It started with John Green submitting an article to McLean's magazine. The article was not used, but was seen by another writer who reworked the material for a story in The Standard. A copy of this story was sent by someone to Ivan Sanderson, causing him to visit Willow Creek, California on his U.S. tour. This eventuality led to Sanderson's article in **True Magazine** (December, 1959) which sparked Roger Patterson's interest in sasquatch.

René and Barbara Wasson Butler, 1995. Barbara was also a sasquatch enthusiast who entered the field in 1966. René and Barbara were great friends in the 1970s and traveled together on bigfoot investigations. Barbara wrote a book, **Sasquatch Apparitions,** that was published in 1979.

May 1995 "Sasquatch Daze" at Harrison. From left to right, front: Larry Lund, Warren Thompson; center: Dan Perez, René Dahinden, Dan Murphy, Barbara Wasson Butler; back: Robert Milner, John Miles.

In the late 1990s, the 15-inch 1958 Bob Titmus Bluff Creek cast (right foot) found considerable fame in a commercial campaign by the Labatt Brewing Company - Kokanee beer. About 400 of the casts were produced by Labatt (not Dahinden) and used for prizes in draws. Lifesize cardboard images of René as shown here greeted shoppers at numerous liquor stores. René also acted in television commercials for Kokanee beer and actually won an actor's award for his performances. While the casts Labatt produced were exact duplicates of the master cast, they were a little too perfect for my liking and lacked the luster of river sand and other natural imperfections seen in casts duplicated by René.

This framed picture was presented to René by the Kokanee people. It shows René carved into a mountains side along with the dog, "Brew," and the sasquatch (played by William Reiter) seen in Kokanee beer commercials. The idea, of course, is a reflection of Mount Rushmore which shows the heads of four U.S. Presidents carved in the mountain side.

Both a copy of the Bob Titmus 1958 Bluff Creek cast (right foot) and one of Roger Patterson's film site casts were duplicated many times and sold to sasquatch researchers, enthusiasts and museums. I worked with René in marketing the casts. Many casts were also produced and sold by Dr. Grover Krantz through Washington State University. René is seen in the following photograph making a new batch of casts. All casts we sold were identified, autographed and dated on the back.

René autographing a cast.

Your author (left) with René Dahinden in the early days. René passed away on April 18, 2001 at the age of 70 years. I learned a lot from this unusual man and his equally unusual passion. For certain, he left larger footprints in the trails of history than those of the creature he so eagerly sought.

Wanja Twan, René's ex-wife, addresses the Celebration of Life gathering for René. Many of those present gave little talks reflecting on their association with René.

There are mysteries unexplained within the forest deep
And legend lures many a man. In search of a giant
big footed creature, too elusive to keep.

Among these folks there was a man who stood out well,
as a one of a kind rare breed. For 45 years he traveled
the land, following the Sasquatch's lead.

Throwing his backpack over his shoulder, his camera
in hand, Sasquatch researcher Renè Dahinden
forged his life in this wild land.

Renè's well worn journey has moved us all deeply
as a father, a grandfather and a friend dear.
And in the quietest moments you would swear, that you
can hear, his footsteps winding over some mountain path
so very near.
...Michelle Beauregard

In Loving Memory of

Renè Dahinden

Born
August 23rd, 1930
Weggis, Switzerland

Passed Away
April 18th, 2001
Richmond, British Columbia
Age 70 years

Memorial Service
Saturday, April 28th, 2001
at 1:00 p.m.
The Vancouver Gun Club
7340 Sidaway Road
Richmond, British Columbia

Cremation

During one of my many visits with René, he gave me some documents to read. On the back of one document, I later discovered this initialed statement. This was René's "message" to the scientific community on the sasquatch issue. It is well that we remember him for these words and indeed take his message to heart ourselves.

*If you don't know the FACTS your opinion is of No Value.
R.d.*

TOM STEENBURG THE GIANT HUNTER

Thomas Steenburg has been actively involved in sasquatch research since 1978. Up until September 2002 he lived in Alberta and was the main researcher in that province. He had, however, done some extensive research in British Columbia and moved to Mission, B.C. in 2002 for the express purpose of living and doing research in Canada's "sasquatch" province. Tom has military training and is a rugged outdoorsman. He is one of the few sasquatch researchers who continually goes out into the wilderness. He had one encounter with a grizzly bear and considers himself very lucky that he managed to get up a tree in time - suffering only a lower back wound and a clawed packsack.

Tom has thoroughly documented much of his research in three books. He tells me that his most memorable experience was a chance investigation in 1986 of a sighting along the Chilliwack River. While Tom was in Hope, B.C., an elderly man saw the SASQUATCH RESEARCH sign on Tom's Ford Bronco (previous vehicle) and informed him of the sighting which had occurred three days earlier. Tom did some checking and found out the exact location. He learned that an American couple were camping in the area. After doing some fishing in the Chilliwack River, they hung their catch on a tree back at their campsite. They saw a sasquatch take the fish and wander off. Tom searched the entire area and to his amazement found 110 footprints measuring 18-inches/45.7cm long starting near a little creek across the road from the campsite. The prints went along the creek bed for about 40 yards/36.6m, and then they suddenly turned to the right and headed up a very steep hill at about a 45 degree angle. He lost the trail by a rock slide area. He photographed the clearest prints and then made a plaster cast.

Impressed with Tom's recollection of this event, I visited the location with him in August of last year. Although the Chilliwack

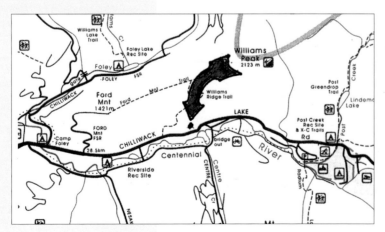

Chilliwack River area 18-inch/45.7cm footprint found by Steenburg and sighting location map.

River area is only some 62 miles/100km from Vancouver, the whole region is heavily forested. There are numerous campgrounds and one can see (unfortunately) that a lot of people use them. We trekked the area of the footprints and although there is now more undergrowth, it is almost incomprehensible how a hoaxer could make prints in the ground that were of the nature Tom found. One can, of course "scuff up" the ground but the results are far from a proper footprint. Nevertheless, even Tom does not overlook the possibility that the prints were fabricated. However, he was not specifically called to investigate the incident as explained and he had to do a fair bit of checking to find the sighting location.

This is the campsite used by the American couple. The road into the area is just beyond the bushes in the background.

*Tom Steenburg and his Chilliwack River area cast which is also shown in the **Footprint and Cast Gallery**.*

The campsite is only a few meters from the Chilliwack River. It was at this spot that the American couple were fishing.

On the Willow Creek, California journey, September 2003. Left to right: Bob Gimlin, John Green, Tom Steenburg, Dmitri Bayanov.

Tom is seen here standing in the path of the footprints he found.

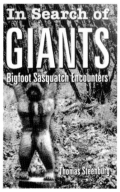

Tom's three books provide highly detail accounts of sasquatch sighting and footprint findings throughout western Canada in recent times. His works include many verbatim interviews with witnesses. (Center and right books published by Hancock House Publishers.)

DANIEL PEREZ THE TIRELESS INVESTIGATOR

"Daniel holds the unique distinction of having been to more alleged bigfoot filming locations than anyone in the world."

Daniel Perez was born and raised in Norwalk, California and became endlessly fascinated by the bigfoot mystery around 1973 after seeing the pseudo-documentary, *The Legend of Boggy Creek*. This documentary triggered his casual to casually serious, to serious fullfledged involvement in this subject matter. Daniel is a union licensed electrician and has been in that occupation since 1985. In 1992, on the heels of the twenty-fifth anniversary of the Patterson-Gimlin film, Daniel authored and published his **BigfooTimes** – *Bigfoot at Bluff Creek*, which is considered by many knowledgeable investigators and researchers as the "bible" on the film. The late René Dahinden once said of the work, "It's the best damn thing ever published on the film." Even today, the slender booklet continues to be a highly authoritative account of the film. It has been used and continues to be used by many researchers and was used for facts in the compilation of this compendium. *Bigfoot at Bluff Creek* went through a second printing in 2003 with 8 pages of bibliography added, changing the work from popular literature to a real piece of scholarship. Although Daniel has researched historical and contemporary cases on bigfoot, he is widely noted for his expertise on the Patterson/Gimlin film - considered by many as the "bedrock of bigfooting." Daniel holds the unique distinction of having been to more alleged bigfoot filming locations than anyone in the world. His research has taken him all over the United States, and also to Canada, Mexico, Russia and Australia. Having investigated and researched the bigfoot mystery for over two decades now, he is completely satisfied that bigfoot is a biological reality, far beyond the realm of fancy and mythology. He writes and publishes his monthly newsletter, **Bigfoot Times**, which is a concise and a factual release. It is a newsletter that addressed areas on the subject

matter that other editors simply would not dare to touch. The publication continues to be the leading newsletter in its class. Daniel also authored the 1988 bibliographical work on the subject, *Big Footnotes: A Comprehensive Bibliography Concerning Bigfoot, The Abominable Snowmen and Related Beings*, which is now considered a standard and requisite reference in the field. He continues to do "on site" investigations of bigfoot sighting and footprint finds and has researched cases in British Columbia, Canada and as far east as Ohio. He is considered a nonsense, extremely serious, meticulous and factual researcher in all matters related to bigfoot studies. Daniel has worked with me and many other researchers on the analysis of the Patterson/Gimlin film, including John Green, the late Rene Dahinden and Dr. Grover Krantz, Dmitri Bayanov and Igor Bourtsev.

To the best of our knowledge, Daniel Perez was the first person to acquire longitude/latitude readings along with an elevation reading of the Patterson/Gimlin film site using GPS technology. It is his opinion that the subject in the film is North America's legendary sasquatch or bigfoot creature.

Daniel and company inspecting the Bluff Creek film site, 2003.

At the Patterson/Gimlin film site, September 2003: From left to right, Dr. Jeff Meldrum, Bob Gimlin, Dr. John Bindernagel, Daniel Perez, John Green, Dmitri Bayanov and Matt Moneymaker.

RICHARD NOLL - A FOREMOST FIELD RESEARCHER

Richard (Rick) Noll has been researching the sasquatch phenomenon since 1969, when he took a vacation trip into the Bluff Creek, California area with a relative working on new bridges in that region. Here Rick became acquainted with the numerous local sasquatch or bigfoot sightings and stories and took up a personal challenge to prove or disprove the creature's existence.

Straight out of high school, Rick served in the U.S. Coast Guard as a sonar technician. An avid outdoorsman, he later trained as a forester at Green River Community College. He is now an expert in aircraft metrology, working for major aerospace manufacturers. He uses such technologies as Theodolite, Laser Tracker, CMM and Photogrammetry. He has published several training manuals on the use of this equipment and provided numerous related lectures. His teaching assignments have taken him around the world, affording many opportunities to check into local reports of sasquatch-like creatures.

Over the last thirty years Rick has worked with all of the major sasquatch researchers and connected with many highly noted anthropologists such as Dr. George Schaller and Dr. Jane Goodall. Rick is seen here with Dr. Goodall in November 2003. Dr. Goodall has expressed belief in sasquatch and Rick interviewed her at that time for a video segment. It should be noted that Rick has become highly well-known not only for his sasquatch related knowledge but also for his expertise in digital technology and related electronics.

"...I prefer the name sasquatch and I firmly believe the animal to be of flesh and blood and living in the Pacific Northwest."

(Media Interview, March 6, 2001)

Rick first saw sasquatch tracks first-hand in 1975 just outside of Twisp, Washington. He and his partner, David Smith, traveled to this area in response to a sasquatch encounter report. The tracks, which were in 2-foot/61cm deep, crusted-over snow, were clearly defined and the encounter report was, in Rick's on words, "hair-raising."

Rick works almost exclusively within his own home state of Washington. He regularly attends sasquatch symposiums (was present at the University of B.C conference in 1978) and provides presentations himself on his sasquatch research and findings. He spends most of his sasquatch research time in the field monitoring several camera traps and track sites in hopes of getting a fresh lead on the creature.

In the adjacent photograph, Rick (right) is seen with René Dahinden on a rock formations at Harrison Lake, British Columbia in 1996. It was here that the idea of using regular bicycles as camera platforms to run old logging roads was conceived. René became so enthusiastic with the concept that he wanted to get a bicycle himself.

Rick is currently one of the curators for the Bigfoot Field Researchers Organization (BFRO) and is recognized as a leader in field investigations (especially regarding history research, photography, measurements and impression casting). He was a major contributor in identifying impressions made by the Skookum sasquatch and making the Skookum body cast. He is the custodian of this highly intriguing evidence.

This photograph of Rick "in the field" captures the essence of Pacific Coast "sasquatch country." It is in these rugged rain-drenched forests that Rick and other members of the Bigfoot Field Researchers Organization spend a great deal of time searching for sasquatch evidence. (Photo taken in 2002 at Rick's special study site which he has been monitoring since the 1970s.)

Rick photographing possible sasquatch footprints on a sandy creek shore at his study site.

NORTHERN EXPOSURE - THE DAUNTLESS J. ROBERT ALLEY

Robert Alley is a sasquatch researcher and author who has been active in the sasquatch field since 1970. As an anthropology student in Winnipeg, Manitoba he communicated with researchers such as Ivan T. Sanderson and Professor Pei of Beijing, China on the possible relationship between the sasquatch and Gigantopithecus. Upon examining Canadian First Nations ethnographics (Labrador to British Columbia) on their belief in hair covered hominids, he concluded that such beliefs were wide spread and generally matched the commonly reported description of sasquatch. He wrote his graduate Bachelor of Arts thesis on this subject. As an undergraduate, he studied zoology and worked as a fish and wildlife technician in the forests of Manitoba, gaining first hand wilderness experience.

Alley went on to study rehabilitation medicine in Manitoba and is well-versed in human anatomy, especially aspects of locomotion. While completing his clinical studies in 1973 he interned at hospitals in British Columbia and Alberta. He took these opportunities to meet the noted sasquatch researchers René Dahinden and Professor Vladimir Markotic (University of Calgary).

During a backpacking trip in the Rocky Mountains, Alley saw possible sasquatch tracks in an area near Nordegg, Alberta. This experience greatly heightened his interest in sasquatch research. While working as a traveling therapist in British Columbia and Alberta (1974/75), he interviewed many people in both provinces on their sasquatch sightings and related experiences. Alley worked directly with Dahinden in investigating several sasquatch sightings in Washington State during that time.

During August 1975, Alley had a first hand encounter with a sasquatch in Strathcona Park, Vancouver Island, B.C. He and three friends had driven up an old logging trail on the east side of Buttle Lake. The group camped at about 2,500 feet/762m in a second growth forest. It was a beautiful summer day with no logging operations or other distractions to disturb the silence and solitude of the great outdoors. While taking in the beautiful view of Buttle Lake far below, the group heard knocking sounds like a baseball bat striking a tree or log. The sounds stopped and then repeated about five minutes later. The group was puzzled with the sounds but soon forgot them and went about preparing dinner.

> "He shone his penlight towards the dark mass of hemlocks and saw a tall black human-like form (like a burnt stump) standing at the edge of the trees."

That night, the group turned-in at about 10:30 p.m. and were very soon fast asleep in their closely arranged tents. Around 1:00 a.m., Alley was awakened by what sounded like a piece of gravel dropping off the side of his tent. He peered out his tent screen and heard his friends peacefully sleeping. As he watched for wildlife near their campsite, something again apparently struck and rolled off the side of his tent. The same thing occurred again and three more times over a 20 minute period with no other unusual sounds. Alley grabbed his penlight and ventured outside, half thinking some teenagers had wandered upon the campsite and were playing a prank. He shone his penlight towards the dark mass of hemlocks and saw a tall (over 6-

feet/1.8m) black human-like form (like a burnt stump), standing at the edge of the trees. The creature immediately charged off into the second growth and in a few seconds was no longer visible. The commotion woke his friends who quickly assembled and asked what was happening. Alley calmly told them he had disturbed an elk while attending needs, rather than upset them with what he had actually seen.

Later in 1975, Alley spent time with René Dahinden, closely studying the Patterson/Gimlin film by viewing it frame by frame. He confirmed Dahinden's impression that the creature's fluid gait was not human in nature and noted the brief prominence of hamstrings and alternate back muscle contractions at precisely the correct moments in the gait cycle. Alley reasoned that a fur suit of any type would not show such details. Dahinden took encouragement from Alley's observations to get the film studied further by scientists.

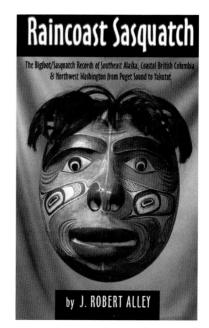

Published 2003 by Hancock House Publishers.

Alley moved to Toronto in 1979 where he graduated with a doctor's degree in chiropractic in 1983. He returned to the West each summer. Between termwork in clinics and hospitals over a period of years, he traveled the length and breadth of both British Columbia and Alberta. He practiced in over 20 different locations (Canada and the United States) where he continued researching local reports of sasquatch sightings.

To maintain a firm grasp on anthropology, Alley took a course in archaeology at Simon Fraser University (British Columbia) and the University of Calgary (Alberta). In Calgary he met sasquatch researcher Tom Steenburg (now of Mission B.C.) and carried out several field investigations with him. Alley moved to the United States in 1988 where he worked as an ergonomics consultant and rehabilitation clinician in the Blue Mountain area of Oregon and Washington. In these states he studied tracking techniques and possible sasquatch signs with the local groups which included the researchers Wes Summerlin, Paul Freeman and Vance Orchard. Alley conferred regularly with Dr. Grover Krantz in Pullman, Washington. In 1991, Alley married and he and his wife, Carol, moved to Ketchikan, Alaska. He has traveled extensively in this state, carrying out exhaustive sasquatch research. His book, *Raincoast Sasquatch* was published by Hancock House in 2003. A natural artist, Alley shows many illustrations in his book that capture the moment described by witnesses. His skillful illustrations have appeared in several books about the sasquatch.

Alley theorizes that there are two, perhaps three, types of sasquatch creatures in Alaska. Reports of sasquatch sightings and sightings of other man-sized creatures come in regularly from all over this state. There are also reports from islands in southeast Alaska (and elsewhere) of alleged bones of a particularly small type of unidentified hair-covered creature.

Besides investigating evidence of sasquatch, Alley enjoys beachcombing with his family and is continuing postgraduate clinical studies on the human foot and human locomotion. Alley pursues studies in ergonomics, injury prevention and forensic anthropology, specializing in gender and age differences in human tracks and gait. He enjoys answering questions about "cryptohominids" and may be reached through Hancock House Publishers, Surrey, British Columbia.

Rob's illustration of a sasquatch reported by a bear hunter James Nunez, at head of George Inlet, Revella Island, Alaska, 2000.

RAY CROWE AND THE INTERNATIONAL BIGFOOT SOCIETY

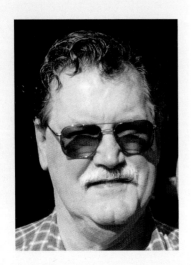

Ray Crowe originated the Western Bigfoot Society in 1991. He has since expanded the operation and now has the International Bigfoot Society as a subsidiary. He runs the organization from his home in Hillsboro, Oregon. Ray conducts monthly society meetings with guest speakers and puts out a monthly newsletter, ***The Track Record***, that contains pertinent information on all aspects of the sasquatch/bigfoot issue. Ray has a remarkable collection of sasquatch related items which he exhibits at his annual sasquatch/bigfoot conference. Speakers from all over North America, and sometimes from Europe, give formal presentations at Ray's conferences.

Ray is a prime contact in the sasquatch field and has participated in television documentaries and radio shows. His opinion is sought by the news media on current issues and he is often quoted in this connection.

Ray is very liberal and fair in his dealings with people. While personally straightforward and reserved when it comes to sasquatch related evidence, at his conferences and in ***The Track Record*** he allows people to express themselves without restrictions. He receives a very high volume of correspondence and he reports on virtually everything allowing readers to judge for themselves.

Rays response as to how he got involved in the bigfoot issue

"An Indian friend told me of their existance and I went into the field with him several times until convinced."

Ray's newsletter The Track Record, as seen here, is a primary publication in the Pacific Northwest. It is both entertaining and informative.

Ray Crowe (left) and your author at the society's 2002 conference. The conferences are well-attended and always get media coverage.

Ray Crowe's collection stretches over many tables. What is shown here is just one section. He has sasquatch hair samples, feces, genuine and hoaxed footprint casts, statues, comic books, rare magazines and more. In essence, it is a "bigfoot without borders" collection. Ray always warns people to put on their "skepticals."

Broken and twisted tree branches found "up high" have been associated with sasquatch. It is thought that they may be markers of some sort. Many researchers, however, don't put much stock in this evidence.

Ray Crowe (standing) coordinates proceedings at his Carson, Washington gathering in 1996.

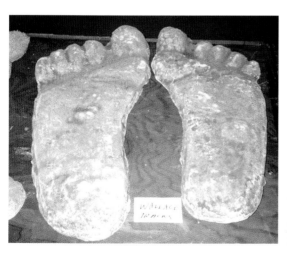

Fake casts produced from faked footprints made with Ray Wallace's infamous wooden feet. Wallace may have fooled some people but he was never a concern to serious researchers - just a nuisance.

MATT MONEYMAKER AND THE BIGFOOT FIELD RESEARCHERS ORGANIZATION

The Bigfoot Field Researchers Organization (BFRO) was established by Matt Moneymaker in 1995. Up to this time, Matt had been involved in bigfoot research for about ten years. With the mass of bigfoot related evidence continuing to come to light as a result of electronic communications, Matt realized the need for a central clearing house to process information. He thereupon designed a proper website and arranged for the necessary professional resources to evaluate findings. To his credit, we now have an organization that has the proper credentials to make decision on all aspects of bigfoot evidence.

The BFRO itself is an international non-profit internet based organization with the primary goal of establishing reasonably conclusive evidence on the existence of bigfoot creatures.

The organization collects and systematically analyses bigfoot sighting reports and other evidence, both submitted by non-members and collected directly by its own field researchers. Members of the organization and other resources include highly experienced researchers, scientists and forensic experts. In essence, the organization is a network of people interested in resolving the bigfoot phenomenon *who are willing to provide their time and knowledge in that quest.*

The BFRO uses its continually updated website to:
• Educate the general public on the history of the bigfoot issue
• Collect information on bigfoot related incidents
• Report the latest findings

The processes the BFRO uses to verify new information or other evidence is highly professional and extensive. It calls on its members in sighting locations to investigate incidents first-hand and report back to the organization. If no member is registered or available in a particular sighting location, the BFRO calls for volunteers in other areas to go to the site.

As a result of the qualifications and experience of its members (and other accessible resources) coupled with proper organization and efficient administration, the BFRO is widely considered the most credible and reliable organization in the field of bigfoot studies.

Members of the organization anticipate that the emphasis on cooperation and professionalism is not only the most realistic approach to resolving the bigfoot mystery, but that it furthers attainment of the BFRO's long term goal: ***The determination of how these rare and elusive animals can and should be protected and studied after their existence is generally acknowledged by governmental agencies and the scientific community.***

Matt's conclusions on bigfoot sightings

"The patterns among eyewitnesses are not demographic, they are geographic - they are not reported by certain types of people, rather by people who venture into certain areas. This simple pattern suggests an external cause."

The BFRO logo has become a symbol of professionalism in the bigfoot field of studies.(www.bfro.net)

BOBBIE SHORT - THE LADY WITH A MISSION

Bobbie Short is a registered nurse. She graduated from the nursing school at Baylor Medical Center, Dallas Texas. Now retired, during her working career she performed private duty in home care with the terminally ill and geriatric patients.

Bobbie became interested in sasquatch in September 1985 after a close sighting of the creature in Northern California. After spending a day backpacking with friends, Bobbie awakened early in the morning and wandered off from the campsite to find a bush away from the sleeping group. From the corner of her eye she saw something moving, heading uphill through a fern field. She stood up to fasten her Levis and looked, in complete disbelief, directly at a large life form walking on two legs and covered from head to toe in a black pelage. It strode quietly passed her and on up the hill in a matter-of-fact but deliberate pace. The creature was close enough at one point for her to see its eyelashes and she noted above all else the look about its gentle eyes. The creature glanced her way for a split second, giving outward recognition of her presence but it never broke stride. There was a "dejected" slump to the creature's upper torso and while it moved extraordinarily fast, the gait seemed awkward if not contorted.

After her remarkable experience, Bobbie dedicated considerable time researching the world's unrecognized primates. She has been on personal missions from the California-Mexican border to the inlets of Wrangell, Alaska searching for information. In 1999, she visited the Pacific Rim countries from the Philippines southward covering a number of islands, including Eastern Samar, Borneo, Malaysia, Negros, and Mindanao.

"The creature was close enough to her at one point for her to see its eyelashes and she noted above all else the look about its gentle eyes."

In 1997, when it was alleged that John Chambers had been involved in the Patterson/Gimlin film, Bobbie visited Chambers and recorded an interview with him. Chambers stated that he was absolutely not involved in the Patterson/Gimlin film and prior to the film had never even heard of Roger Patterson. During the interview, Chambers related on record that he was considered the best movie costume designer, but he definitely could not recreate the creature seen in the Patterson/Gimlin film. Moreover, he declared that John Landis, Baker and others were incorrect in their assumption that he (Chambers) was capable of such a creation, especially in 1967. Chamber's best work was seen in the movie *Planet of the Apes,* which was released in 1968. This movie

John Chambers is seen here in a photograph taken by Bobbie Short when she interviewed him (October 26, 1997) relative to the Patterson/Gimlin film. He was in a nursing home in Los Angeles at the time. Chambers had stated over one year earlier that he did not "design the costume," but this was interpreted by sceptics to possibly mean that he had someone else design it and he fabricated it. Bobbie Short set the record straight once and for all and the matter is now closed with serious researchers.

naturally raised speculation as to Chamber's involvement in the Patterson/Gimlin film. Bobbie, however, has confirmed beyond a reasonable doubt that there was no such involvement.

In 2003, Bobbie covered Central Asia, traveling mostly on foot throughout the Shennongjia Mountain region gathering information on the Chinese wildman, Tibetan Yeti and Mongolian Almas. This was the most productive trek of her research career, enabling her to get a better understanding of mystery primates and other unusual fauna in Communist China.

Bobbie created her website, **Bigfoot Encounters** in September 1996. She moderated the Internet Virtual Bigfoot Conference (IVBC) after Henry Franzoni left and is now editor of her own regular email newsletter which is sent to some 1,410 subscribers world-wide. Her website contains a virtual archive of sasquatch, bigfoot, yeti, wildman and orang pendek related information and documents (magazine articles, newspaper articles and reports). In 1998 Bobbie was one of the original board members of the North American Science Institute (NASI).

Bobbie regularly does her own style of fieldwork, researches new information and informs her subscribers of new findings either directly or by providing links to where information can be found. She writes articles on the subject of mystery primates, maintains a database of information and does presentation for high school children.

She receives many email enquiries related to the sasquatch and either responds directly or redirects people as to where they may obtain required information. Bobbie devotes most of her time to field work and writing her books for future generations to absorb - upholding a stance that sasquatch research isn't a belief system, but an ongoing investigation. In her own words, "I like citations and source information; it keeps Sasquatch research legitimate and accountable." At this writing, Bobbie is scheduled to do a presentation at the Fifth European Symposium of Cryptozoology, to be held in Belgium May 29 to 31, 2004. Bobbie will be speaking on hirsute hominids in China, Mongolia and Tibet and will present tracks found in the Pacific Rim Countries of Borneo and Malaysia. Bobbie's website is at: http//www.bigfootencounters.com.

PAUL SMITH - THE ARTISTIC VISIONARY

Paul Smith is a fine arts artist, illustrator, graphic designer and teacher living in Seattle, Washington. Paul has been creating artistic images for many years. Upon entering the sasquatch/bigfoot field, Paul concentrated on creating images of the creatures in their various walks of life.

I have presented here three of Paul's works to both provide the reader with some insights into this aspect of study and also to illustrate the creatures in their natural environment. Paul has based his imagery on historical and present descriptions of the creature together with situations and creature activities reported by witnesses or reasonably known to have occurred. Artists of Paul's calibre have remarkable insights which very often only need to be confirmed by a camera.

*Our immediate impression when hearing the word "sasquatch" or "bigfoot" is of a male "rogue" creature. However, to have such, there needs to be females and babies to begin with. Here we have what might be a typical sasquatch family. Paul Smith's art work is available on line at **paul.smith7@comcast.net**.*

Here we see a sasquatch foraging for food or perhaps obtaining bark to make a "nest" or shelter. The forests of North America sustain many large creatures. There is little doubt that the sasquatch could both survive and propagate in these regions.

*At best, most sightings of sasquatch are little more than a fleeting glance - no time for cameras. Here, Paul captures one of those moments which many of us long to experience. Paul Smith's art work is available on line at **paul.smith7@comcast.net**.*

Intriguing Associations

THE RUSSIAN SNOWMAN

For centuries, there have been sightings and stories of unusual hominid creatures in Russia. Such creatures are depicted in early Russian drawings, paintings sculptures and engravings. While the creatures share many of the same features as the North American sasquatch or bigfoot, they are not considered the same species. Dr. Grover Krantz sums up his appraisal as follows:

Most of the Caucasus descriptions could be fitted into a sasquatch mold, but only with considerable difficulty. The size, and especially the massiveness, of the sasquatch body is not evident here, though it could be a geographical variant in this regard. More problematical is the notable sexual dimorphism of the sasquatch and its distinctively different (and much larger) footprints. The non-opposed thumb is like the sasquatch, but the elongated fingers are not. Its behavior is also somewhat different, especially in its interactions with humans and their dogs.

As to my knowledge there are no photographs of a Russian Snowman, it is apparent Dr. Krantz has used verbal descriptions of the creature for information regarding the thumb and fingers. The photographs shown here are those of alleged Russian Snowman footprints. The first print, which measured about 15.5-inches (39.4 cm) in length, was found in Tien Shan in 1962. One thing that strikes me about this print is the position and comparative insignificance of the little toe. Is it possible this toe might fail to make a significant impression in some prints, giving rise to what might be conceived to be a four-toed print? Four-toed prints have been found in both Russia and

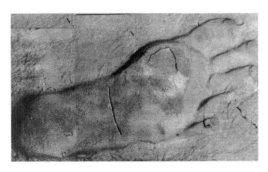

North America. The next print, one in a series about 10-inches (25.4 cm) long, was found in March 1978 in the Dolina Narzanov Valley, North Caucasus.

Research in Russia is carried out mainly by the Relict Hominoid Research Seminar, organized at the Darwin Museum in Moscow. The chairman, Dmitri Bayanov, after extensive field-work, authored a book, *In the Footsteps of the Russian Snowman,* published in 1996 (Crypto-Logos Publishers). This book provides very convincing evidence of the creature's existence, giving us remarkable eyewitness accounts. The major and most fascinating accounts provided are those concerning what I call the "Karapetian Hominid" and the story of Zana. Details on each of these creatures follow. The photograph seen here shows Dmitri Bayanov in the hills of Kabarda, North Caucasus in the 1970s.

Lt. Col. V.S. Karapetian, MD

THE KARAPETIAN HOMINID: We are told that in December 1941, a Russian army unit in the Caucasus observed a strange hairy man near their post. Fearing that he might be with the enemy, soldiers quickly captured him. Because of the man's unusual appearance, Lt. Col. V.S. Karapetian, a medical doctor in the Army Medical Corps, was asked to examine him. The following is Dr. Karapetian's statement made to a magazine correspondent on the incident:

The man I saw is quite clear in my memory as if standing in front of me now. I was inspecting him on the request of local authorities. It was necessary to establish whether the strange man was an enemy saboteur in disguise. But it was a totally wild creature, almost fully covered with dark brown hair resembling a bear's fur, without a mustache or beard, with just slight hairiness on the face. The man was standing very upright, his arms hanging down. He was higher than medium, about 180 centimeters (71 inches). He was standing like an athlete, his powerful chest put forward. His eyes had an empty, purely animal expression. He did not accept any food or drink. He said nothing and made only inarticulate sounds. I extended my hand to him and even said 'hello.' But he did not respond. After inspection I returned to my unit and never received any further information about the strange creature.

Dr. Karapetian extends his hand to the unusual hairy man. (Drawing by Lydia Bourtseva.)

*Both Dmitri Bayanov and Igor Bourtsev have done extensive research on North America's sasquatch or bigfoot. The books shown here were authored by Dmitri. The first book, **America's Bigfoot: Fact, Not Fiction**, discusses the overwhelming evidence supporting the existence of the creature. The second book, **BIGFOOT: To Kill or To Film; The Problem of Proof**, addresses the troublesome question of our right to kill one of the creatures to prove its existence to the scientific community.*

In providing further details at a later date, Karapetian revealed that the man was cold-resistant, and preferred cold conditions to normal room temperature. He was shown to Karapetian in a cold shed and when the doctor asked why he was kept in such cold conditions, soldiers informed that he had perspired excessively in the building where he was first taken. Elaborating on the man's face, Karapetian stated that the subject had a very non-human, animal-like expression. Moreover, Karapetian revealed that the man had lice of a much larger size and of a different kind than those found on humans. The doctor informed the authorities that the entity was not a man in disguise but a "very, very wild" subject with real hair. It has been generally accepted that the drawing shown here was created by Karapetian. This information is not correct. It is apparently an artist's conception that was created for the story at some point in time and has been incorrectly identified as a drawing made by Karapetian.

ZANA: The story of Zana, a Russian ape-woman, is truly remarkable. Zana died in the 1880's or 1890's so some people in the area she lived actually remembered her when researchers questioned them in 1962. It is believed hunters captured her in the wild whereupon she was sold. She changed hands several times and eventually became the property of a nobleman. The following description of Zana is quoted from Dmitri Bayanov's book, *In the Footsteps of the Russian Snowman*.

Her skin was black, or dark grey, and her whole body covered with reddish-black hair. The hair on her head was tousled and thick, hanging mane-like down her back.

From remembered descriptions given to Mashkovtsev and Porshnev, her face was terrifying; broad, with high cheekbones, flat nose, turned out nostrils, muzzle-like jaws, wide mouth with large teeth, low forehead, and eyes of a reddish tinge. But the most frightening feature was her expression which was purely animal, not human. Sometimes, she would give a spontaneous laugh, baring those big white teeth of hers. The latter were so strong that she easily cracked the hardest walnuts.

Zana was trained to perform simple domestic chores and became pregnant several times by various men. Remarkably, she gave birth to normal human babies, four of whom survived to adulthood (two males and two female). The youngest child, a male named Khwit, died in 1954. All of the children had descendants.

Several expedition were made in the 1960's and 1970's (notably those headed by Professor Boris Porshnev and later Igor Bourtsev) to find Zana's grave and exhume her remains for examination. While many sites were explored, the researchers were unable to find a skeleton that matched the description of Zana. On the Bourtsev expedition of 1978, it was decided to exhume the remains of Khwit whose grave was well indicated. The idea, of course, being to determine what traits he had inherited from his mother. Khwit's skull (seen here) was taken to Moscow where it was studied.

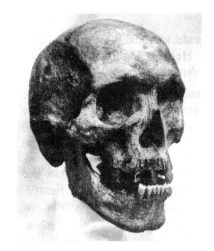

Skull of Khwit.

Igor Bourtsev is shown here examining Khwit's skull at the gravesite. While there was some talk in the 1990's that the skull shown was not, or may not have been, that of Khwit, I have confirmed that this information was totally incorrect. Igor is definitely holding Khwit's skull.

*NOTE: Igor Bourtsev provides us with further information and insights on Zana under **Abkhazia** in his paper which follows.*

While Russian anthropologists reported that the skull was different from that of ordinary human beings, such was not the opinion of Dr. Grover Krantz, an American anthropologist. Krantz stated that the skull is from a fairly normal, modern human. In the adjacent photograph, Dr. Krantz is seen holding Khwit's skull with Igor Bourtsev looking on. The photograph was taken while Dr. Krantz was in Moscow in 1985.

Dr. Krantz is seen holding Khwit's skull with Igor Bourtsev looking on.

HOMINOLOGY IN RUSSIA
PERSONAL OBSERVATIONS, PROBLEMS AND PERSPECTIVES

By Igor Bourtsev

North Caucasus: After the unsuccessful Pamir Expedition of 1958, the North Caucasus became the main region of searches for the "snowman." Marie-Jeanne Koffmann's expedition worked there for decades. Her headquarters was based in the settlement of Sarmakovo on the Malka River in the Soviet Autonomous Republic of Kabardino-Balkaria (in Russia). The creature in question is called there *almasty,* sometimes *kaptar.* I joined Koffmann's expedition in the summer of 1965, which was my first one in that respect. The information collected there within one month was so convincing that the search for hominoids became ever since the

Photo of Khwit.

main goal of my life. In nearly 40 years that have passed since then I changed several professions. But all through this time and whatever positions I occupied, I have always remained a hominologist. At every opportunity I engaged in the search and investigation of relict hominoids. So it is possible to say that my basic and constant specialty is hominology.

Azerbaijan: After my first expedition season of 1965, there was a rather long break in my field work on account of a lung disease which sent me into hospitals and sanatoriums. But I used this enforced illness for the study of anthropological disciplines and foreign languages, English in the first place. Besides, I took an active part in the organization and preparation of Koffmann's expeditions to the North Caucasus, in collecting information from other regions of the Soviet Union, and participated in the work of the Smolin Hominology Seminar at the Darwin Museum. My next expedition to the mountains was in 1970 the Talysh Mountains in the south of Azerbaijan, on the border with Iran. I had read several sighting accounts from that area in Boris Porshnev's book and wanted to ascertain something, and if lucky, at least find hominoid tracks. I had also been given a tip by the chief teacher of a school in Baku (capital of Azerbaijan), named Pinya Kalika. He had been on excursions with his pupils in the Talysh Mountains and had heard local people talk about "wild men." He gave me addresses of the locals who could help me and even serve as guides. I traveled there with my previous wife, Alexandra Bourtseva.

> "They even had to fire their rifles in the air when the creature came too close."

I managed to establish friendly relations with the local people in the Talysh mountains (they called themselves Talyshs). Their population is about one hundred thousand; their language is different to Azerbaijanian - close to Persian. There I started to study their language and some of them entrusted their stories to me about the mysterious wild bipeds. In Azerbaijan they are called *guleibany* (probably derived from the Arabic "*goul'*) or *meshe-adam*, i.e., wood man). But the Talysh people have names of their own for these creatures, *alazhen* for females and *alamerd* for males. In the course of the Talysh expedition, we collected about 30 accounts concerning hominoid sightings and encounters which not always had a peaceful ending. Most memorable are two cases. The first happened in 1948, during the time of local elections. At night three men, members of the election commission, were carrying the ballot box from a village to the district's election center. All along the way they were followed by a wild biped two and half meters (8 feet) tall. They even had to fire their rifles in the air when the creature came too close.

Another case occurred also in the 1940s and concerned a militiamen (i.e., policeman), named Shahbala Huseinov. Once he

218

was returning after duty late at night, and on a bridge that crossed a creek he was attacked by a couple of homins, a female and a male. The female was taller than the male but the latter was stouter. The female knocked the militiaman down; he fell on his back and was afraid to move. She touched bright buttons on his uniform and mumbled, while the male growled in displeasure. They lost interest in him in about an hour and went away. The militiaman got up and trudged home without his cap which fell off when he was knocked down. Only then did he remember that he was armed with a pistol. After that he was ill for a long time. He told me that story himself and showed me the place where this happened to him.

"The female was taller than the male but the latter was stouter."

I spent five expedition seasons in Talysh, 1970-1975. Aside from eyewitness accounts, I did not obtain any other evidence of the homins' presence in that area. At Talysh, I studied the "braided" manes of horses. The locals claim that the braiding is the work of "wild men." I cut off some 40 braids from different horses. I investigated the phenomenon by watching grazing horses at night through a night vision device. Then I understood that the braids could appear by themselves. This can happen when something like a piece of clay or a thorn fastens to the ends of three tresses of a horse's mane. When the horse shakes its neck, the piece of clay sometimes gets into one gap between two tresses, sometimes into another gap, and thus the hair becomes braided from the ends of hair tresses to the roots. I concluded that at least some braids originate by themselves (this story is provided with photographs in Dmitri Bayanov's book, *In the Footsteps of the Russian Snowman*.) After that I made no more expeditions to the Talysh mountains.

During my expeditions in Talysh, I learned that a well-known Georgian zoologist and palaeontologist, Nikolai Burchak-Abramovich, was also searching for homins in that area. I was told that he had found their tracks and made casts of them. Professor Burchak-Abramovich discovered (in the Caucasus) the tooth of a fossil primate, *Dryopithecus*, named by him *Udabnopithecus*. Later on, Burchak-Abramovich helped me in my searches for the grave of Zana in Abkhazia.

Farmers showing the braided manes of their horses, said to be the work of "wild men."

Since January 1972, after René Dahinden's visit, I concentrated on the study of the Patterson/Gimlin film and this investigation has continued until now. The more I study this film the more I am convinced that it is genuine. I will not dwell on this subject here because it is fully presented in other publications.

Abkhazia: The story of Zana is that of a wild woman, caught and habituated by people in Abkhazia at the end of the 19th Century. She not only lived with people but was the mother of children by human fathers. This information is in Boris Porshnev's book and at the time I read it I was not only acquainted with Porshnev but was also helping him in his research. My subsequent research on Zana is provided in detail in the above mentioned book by Bayanov. Here, I only want to emphasize three points.

First, the story of Zana is not simply a fascinating tale about a surprising contact of people with a wild man-like creature. This story is one of a number of episodes remarkable from the point of view of the theory of parallel existence of Homo sapiens and non-sapient hominids, and their crossbreeding throughout the course of history. There are other cases of probable crossbreeding of this kind as, for example, in the Sungir excavation (of twenty-three thousand year antiquity) in central Russia, where, in a sapiense burial, were found bones with Neandertal features. The Zana case could shed more light on this problem. The point is we have the skull of Zana's son, Khwit. Besides, Zana's descendants live in Abkhazia and DNA analyses of their blood could be of much help in verifying the Zana story. During one of my trips to Abkhazia, I obtained blood for analysis from the daughter of Zana's son - that is a granddaughter of Zana. Unfortunately the result of that analysis is unknown to me because of the ethnic war and conflict that engulfed Abkhazia - which is still not over and prevents our further research in that area.

Igor Bourtsev in search of Zana's remains. Bourtsev headed three expeditions in this quest but never found the wild woman's skeleton.

Second, I dug out an unusual burial of a female near the grave of Zana's son Khwit. The rubber footwear of the buried female had the date of its make - 1880. Khwit was born, according to his documents, in 1886. But considering the fact that personal documents began to be issued in Abkhazia only in the 1930s, on the basis of oral statements, we can suppose that Khwit was born earlier. On the other hand, footwear could have been kept for many years after it was made - such a custom exists in those parts even today. The study of the female's skull showed its Negroid features, whereas Khwit, according to anthropologists, looked very much Australoid. I learned that some Africans lived in Abkhazia in the 19th century, and even found several of their descendants. But all those who had seen Zana insisted that she was not African - her body was covered with hair and she remained wild, despite all attempts to civilize her.

I personally restored the female's skull at the Laboratory of Plastic Reconstruction, headed after Mikhail Gerasimov's death, by anthropologist Galina Lebedeva. She consulted me during the restoration and highly praised the quality of my work. She also made a drawing of the woman's face, based on her skull, and that portrait clearly shows African features. At the same laboratory, Khwit's skull was examined for the absence or presence of a pathological condition, called acromegalia - and it was found that no pathology was present. Modern methods allow us to determine by analysis of bones whether these two skulls belong to relatives. If yes, then the buried female was the mother of Khwit, that is Zana. If not, then the female is clearly not Zana. At present I believe the second version. But for 100% certainty it is necessary to undertake a corresponding analysis. Unfortunately, we have no funds for that and have not been able so far to interest those who have.

Third, the process of excavation was connected with some mysterious phenomena: an unexpected strong rain storm when finishing the excavation; my personal severe illness with unknown diagnosis that followed two or three days later; the "uncomfortable feelings" of a person with ESP ability who later visited the burial place and so forth.

Mongolia: In 1976 we showed the Patterson/Gimlin film to a prominent Russian academician, archaeologist Dr. Alexey Okladnikov. At the time he headed a major Soviet-Mongolian archaeological and ethnographic expedition in neighboring Mongolia. The film and our talk about the Mongolian almas so much impressed the academician that he immediately decided to organize a group for an expedition to search for almases and he included me in that group. Thus in 1976 I made a fact-finding trip to Mongolia. I learned that the Mongolian prime minister, Maidar, was keenly interested in the almas problem and even devoted a chapter to it in his book about Mongolia. I also met the academician Rinchen, a pioneer of the almas research, but his advanced age did not allow him to be active any longer. In the company of three colleagues I traveled in the western regions of Mongolia. A detailed account of that trip needs a separate report. I can only say here that conditions for fieldwork there are not easy. The landscape is barren and strong cold winds are blowing, but the fauna nonetheless is very rich, especially in marmots, so there is plenty of food for the almases. Unfortunately, academicians Okladnikov and Rinchen died soon after, and the new head of the Soviet-Mongolian expedition was not at all interested in almases. Our thrust, therefore, in that direction had to stop. In those years, or somewhat later, Mongolia was visited by British anthropologist Myra Shackley who published a book about her findings. She wrote about the possible existence of Neanderthals in Mongolia, thus supporting Porshnev's hypothesis. Dmitri Bayanov and I had occasion to meet Myra Shackley in Moscow. I shared with her some of my material and I understand she used some of it in her publications.

Children such as those shown here who live in the vast rural areas of Russia are probably more likely to see an almasty or snowman than members of fully-equipped expeditions to find the creature.

Tajikistan: Expeditions to Tajikistan, led by Igor Tatsl, of Kiev, Ukraine, and I are documented in Bayanov's book. I wish only to draw attention to a particular aspect of the story. During those expeditions we made planned attempts to contact the homins. The best known is the case of Nina Grinyova who volunteered to "date" a homin in an open place and who succeeded. The meeting strongly affected Nina's subconscious; she received, so to speak, a mental shock and became unconscious. It was thanks to the actions of other people that she came-to. Another case in that expedition was an unsuccessful attempt to photograph a homin who approached the tents of a group at night. One of the group, an engineer, knowing of the creatures' ability to affect human mind, prepared his camera and what's more the tent itself against what he presumed to be the electro-magnetic influence of the homin's vibes. He simply screened his tent with foil. Nonetheless, when he heard heavy steps outside and tried to raise himself toward the window with camera in hand his body suddenly felt so heavy that he couldn't even lift his arm. He was sort of paralyzed. Only when the steps moved away was he able to move and look out, but there was nobody in sight.

In that expedition I cast one of the best homin footprints discovered in the Soviet Union *(shown later)*. A set of footprints appeared during the night some twenty metres (70 feet) from our tents, but not all were clear. I made a cast of the best one - 35cm (13.8 inches) long and 15cm (5.9 inches) wide at the ball. I am certain hoaxing was excluded. Another important result of the expeditions was the discovery of scat of such form and size which leave no doubts of its belonging to a big homin.

Expeditions to Tajikistan continued through the 1970s and 80s, with such participants as Dmitri Bayanov, Vadim Makarov, Michael Trachtengerts, Gleb Koval and others. Our work stopped in that region with the collapse of the Soviet Union and the start of military operations in Tajikistan. As for me, I went to work in Afghanistan in 1985 and returned home only in 1988. In 1991 Igor Tatsl died, also the political situation changed in our country. Tajikistan became an independent state, the same as Ukraine from where the expeditions by Tatsl had been organized. Expedition to Pamiro-Alay also stopped.

The political and economic changes in Russia influenced our activities. There was a period when our fieldwork completely ceased because of hard economic conditions. Of late it is picking up, especially after we received news of recent sightings in the Kirov Region in the north-western part of European Russia. I made trips to the area with our younger colleague, Gleb Koval, and we have established that the area has very promising perspectives for further work, both on account of good recent sightings and the positive response and attitude of the local authorities.

Events in the Kirov region have activated sessions of the

> "Another important result of the expeditions was the discovery of scat of such form and size which leave no doubts of its belonging to a big homin."

Nina Grinyova talking to a local eyewitness.

Smolin Hominology Seminar and the Russian Cryptozoology Association. On the initiative of Gleb Koval, we have just established a fund for promotion of scientific investigations and searches, named Cryptosphere. So far there is not a kopeck of money in the fund's account, but we hope to draw attention to our activities and raise money for our work. We see our main goal in establishing friendly contact with the homins for subsequent videotaping and photographing. With these aims in view, we are publishing our newsletter, *The Herald of Hominology*, first only in Russian, but if we find sponsors abroad (in particular, the U.S., Canada and the U.K.) we could prepare an English version - also in other languages on the basis of partnerships. I have already published the first issue - 32 pages (usual sheet size) with photographs and impressive drawings.

Igor Bourtsev
December 10, 2003

Dmitri Bayanov is seen here with his beloved "Patty," (who Dmitri himself named) the creature in the Patterson/Gimlin film. The eight-foot enlargement of Frame 350 of the film was displayed at the Willow Creek Bigfoot Symposium in September 2003. Over many years Dmitri has talked to me with such endearment for the creature I jumped at the opportunity to bring the two together.

From left to right, Dmitri Bayanov, Dr. Dmitri Donskoy and Igor Bourtsev are seen here in about 1976. The statues shown were created by Igor Bourtsev of the creature seen in the Patterson/Gimlin film. Igor is holding a photograph of the Russian Snowman print found in Tien Shan Mountains in 1962 (previously illustrated).

Igor Bourtsev (left) and Dmitri Bayanov are seen here in a photograph taken by Daniel Perez on his visit to Russia in 2002. Igor and Dmitri have spent many years in the field of hominology. Their painstaking contributions have greatly influenced acceptance in some sectors of the general scientific community as to the reality of living hominid creatures.

The founders of hominoid research in Russia. From left to right: Boris Porshnev, Alexander Mashkovtsev, Pyotr Smolin, Dmitri Bayanov and Marie-Jeanne Koffmann. (Photograph taken in January 1968).

This remnant of a possible snowman shelter was found in the Kirov region of Russia (near the Urals) in 2003. People in this region have lately reported considerable "wild men" activity and local authorities are assisting in the search for the creature. Igor Bourtsev has been to the area twice in the last two years. He informs us that local researchers are now using automatic video cameras along the supposed paths of the creature.

Igor Bourtsev compares his foot to a cast made from a footprint found in the Pamir-Alai Mountains (Tajikistan Republic), August 29, 1979. Several footprints were found in the morning about 70-feet/21.3m away from his group's tents. The cast is about 14-inches /35.6 cm long.

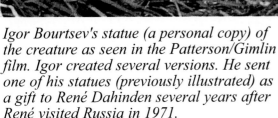

Igor Bourtsev's statue (a personal copy) of the creature as seen in the Patterson/Gimlin film. Igor created several versions. He sent one of his statues (previously illustrated) as a gift to René Dahinden several years after René visited Russia in 1971.

This photograph shows Igor Bourtsev on the right with two Kirov region huntsmen. The center huntsman, Valery Sergeev, states that he has met wild men and their women and children many times over the last twenty years. (Photograph taken in 2003)

THE YETI

The yeti, which is said to inhabit the Himalayan Mountains, was first brought to the attention of the outside world about 100 years ago. Since that time, many expeditions have been undertaken to find the creature. There are many documented sightings, some very credible, but absolutely no photographic evidence. Alleged footprints in the snow remain the main tangible evidence of the creature's existence. The yeti footprint cast (copy) shown here was created from a photograph. The cast is about 12.5-inches (31.8cm) long.

An alleged yeti scalp (one of three known to exist) was professionally examined and declared to have been made from the skin of a serow, a member of the goat antelope family. It has been concluded that all known scalps are therefore *likely* fabricated.

It is possible, of course, that the scalp examined and other scalps were copied from an *original* yeti scalp. The yeti is held sacred in Tibet so when one monastery *possibly* obtained a real scalp, all other monasteries wanted one. The monks in the other monasteries therefore made duplicate scalps. Over the centuries all scalps would became "real" in the eyes and hearts of the monks. The inference here, therefore, is that hidden away in some lofty secluded monastery rests a real yeti scalp. We might just wonder if the monks who have the original would even allow it to be viewed by outsiders, let alone be taken away for analysis. Is it possible the researchers were sidetracked? Nevertheless, it is entirely possible that one of the two scalps known to exist but not examined is the real scalp. These scalps are about 350 years old. I don't have the specific age of the scalp examined.

Further, an alleged yeti skeletal hand held together with wire was also found and professionally examined. It was declared to be made from part human and part animal bones. However it is now known that a researcher had previously stolen some of the bones from the hand and replaced them with human bones. When the hand was examined, human bones were naturally determined. No specific identification was provided for the "animal" bones. Certainly, another examination should be performed on the hand, but I wonder if we could now get access to it - if it is still there. (I have been informed that the entire hand was stolen in the late 1980's).

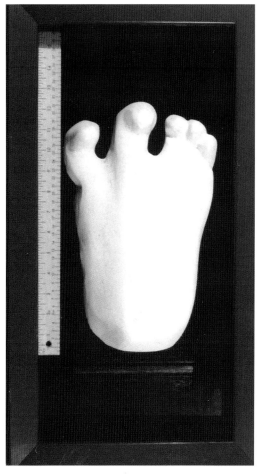

"The inference here, therefore, is that hidden away in some lofty secluded monastery rests a real yeti scalp"

"We have never encountered DNA that we couldn't recognize before."

The latest information on the yeti appeared in ***The Times*** (Britain) on February 4, 2001 and the entire story was later provided in a television documentary (*To the Ends of the Earth* series). A team of British scientist went on an expedition to Bhutan to seek evidence of the yeti's existence. Here they obtained the services of a resident "official yeti hunter." The yeti hunter told the scientist he had seen the creature enter a hollow at the base of a large cedar tree. He then led the scientists on a long arduous trek to the tree which was situated in a forest in eastern Bhutan. One of the scientists, Dr. Rob McCall, a zoologist, obtained hair strands from the entrance to the hollow. It appears the creature scraped its shoulders or upper back against the tree as it bent over to enter the hollow thereby leaving hair strands. The hair was analyzed in Britain by Bryan Sykes, Professor of Human Genetics at the Oxford Institute of Molecular Medicine. Sykes stated: *We found some DNA on it, but we don't know what it is. It's not human, not a bear nor anything else we have so far been able to identify. It's a mystery and I never thought this would end in a mystery. We have never encountered DNA that we couldn't recognize before.*

My only comment here is that it is unfortunate the yeti hunter does not seem to own a camera and I hope the British scientists provided him with one.

Like sasquatch, the yeti has also found postage stamp distinction. In 1966, Bhutan issued stamps showing five different views of the creature on fifteen different stamp denominations. The five designs are shown here. It does not appear, however, that the stamps were an official government issue. Like the creature itself, the stamps are a bit of a mystery.

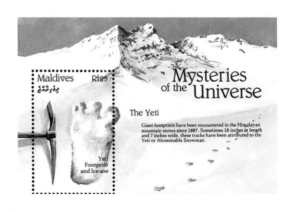

This Maldives Islands stamp souvenir sheet (1992) shows one of the yeti footprints discovered Eric Shipton and Michael Ward on their 1951 Himalayan expedition. The information shown on the sheet reads:

The Yeti: *Giant footprints have been encountered in the Himalayan mountain snows since 1887. Sometime 18 inches in length and 7 inches wide, these tracks have been attributed to the Yeti or Abominable Snowman.*

This is a portrait of a yeti by the famous naturalist artist, Robert Bateman. Bateman would have based the image on numerous eyewitness accounts of the creature and the depictions of other artists. To my knowledge there is no photograph of a yeti. Whatever the case, Bateman's insights are significant. If the creature does exist, then I would venture to say that the likeness shown here would be very close.

11

Conclusion

The evidence presented in this work definitely indicates that sasquatch exist. Certainly, men have been sent to the gallows on the strength of less convincing evidence.

The Patterson/Gimlin film might be considered the main piece of sasquatch evidence and has been referenced as the "gold standard" of sasquatch research. The film, however, is just one piece of evidence. Initially, the film's credibility was primarily based on the other evidence. Certainly, without previous sightings and footprints, the film would not have received the same degree of interest and attention.

Nevertheless, make no mistake; the film can "stand on its own two feet." Consider for a moment a scenario whereby the film was the only evidence. In other words, before and after the filming absolutely no other sasquatch evidence was found (no history references, no sightings, no footprints and so forth). Assume that Patterson and Gimlin just happened to spot and film something unusual which they named a bigfoot. Given the film was subjected to the same degree of analysis, would conclusions reached have been different? The answer is no, absolutely no difference. What would change, of course, would be our concept of the creature's population and range.

All we could say with certainty is that as of the afternoon of October 20, 1967 there existed a bigfoot at Bluff Creek, California. We could only assume the creature, and perhaps others of its kind, existed at that place and elsewhere before and after the filming.

Now, let's look at the other side of the coin. Assume there was no Patterson/Gimlin film. How much impact would this condition have on the *credibility* of all other evidence? The answer is none, absolutely no impact. Numerous sightings and footprint finds predate the film. Further, some people who report sasquatch related incidents may not even be aware of the film. Certainly, these conditions do not detract from the credibility of their sightings or findings.

So what would change in the whole sasquatch issue? Mainly, we would not have reasonably clear photographs of what *one* of the creatures looks like, its measurements and that creature's walking pattern and rate (these latter points, incidentally, serve to confirm the reality of the filmed creature). Nevertheless, absence of the film would reduce *support* for the other evidence. It is one thing to have sightings and footprints and quite another to have this evidence and a photograph of the creature.

While there are other photographs beside the Patterson/Gimlin film, we have not been able to establish their authenticity. If the question is: To what degree would absence of the film reduce support? The answer is: For the press people and the general public, significantly; for serious sasquatch researchers, marginally or not at all.

In some ways, and I say this without reservation, we would have been further ahead in sasquatch research if the film did not exist. Too much time, energy and resources has gone into defending the film and disputing what it does or does not reveal. The mistaken "notion" that disproving the film (proving that it is a hoax) automatically disproves sasquatch, or at least strikes a major blow to sasquatch researchers, presents a challenge and opportunity for notoriety. I have pointed out why this notion is incorrect.

In essence, the film provided a "target" on which negative *circumstantial* evidence and perceptions can be brought to bear. As such the film has become a "weak spot" as well as a "strong spot" in the entire sasquatch issue. Indeed, when the film was first viewed by scientists at the University of British Columbia and elsewhere, perceptions were formed that cast a shadow on all evidence of the creature's existence. We are still trying to remove that shadow.

In summary, the main value of the Patterson/Gimlin film is that it supports all of the other evidence we have. The film is not needed to validate the other evidence. The other evidence stands alone in this regard.

Given that sasquatch do exist and are found to be more closely related to human beings than any other primate, what would be the impact? On this question, Professor Boris Porshnev stated: ***I see the situation as a scientific revolution;*** Dr. John Napier stated: ***We shall have to rewrite the story of human evolution;*** Dr. George Schaller stated: ***Finding another species that's more closely related than any other to human beings would have a tremendous impact on humanity.***

If the creature is found to be no more closely related to human beings than other primates, then we have a new animal species - a North American ape. While this finding would not be as significant as a "more closely related species," it would still be a remarkable discovery. Regardless of the nature of the creature, one important question will certainly be asked: If sasquatch are "out there" what else might there be? In the same way as the Patterson/Gimlin film supported sasquatch, this creature itself will now lend support to the existence of other creatures in the annals of cryptozoology.

"I have met Bob Gimlin (left) and talked with him at length about the filming of the creature at Bluff Creek. Anyone who would have the pleasure of meeting Bob would be as convinced as I am as to his honesty and sincerity."

Chris Murphy

A CALL TO ACTION

It has now been over thirty-six years since Patterson and Gimlin provided us with photographic evidence of a sasquatch. Since then there have been other photographs (although not as convincing), thousands of sightings (about 400 a year), hundreds of footprints and several pieces of other evidence "placed on the table." *Nevertheless, we are still at "square one" in getting a government (or major research organization) sponsored investigation to find the creature and then effecting legislation to protect it and its habitat.*

Naturally, the first step in the process is to convince the general scientific community that the creature exists. In other words, get most scientists to agree that the creature is "out there." After that, we would simply lobby for appropriate action. Here, however, we meet with the obstacle of having to provide a "body or bones." Ironically, if we had a body or bones we would not need an investigation to prove the creature exists. With scientists it is the old case of, "I'll see it when I believe it."

Nevertheless, to say we have not progressed in getting scientific recognition of the creature would be incorrect. There are now many eminent scientists involved in serious research, but it is a very slow process. Certainly if some very high profile scientist or naturalist were to actually see one of the creatures then the process would move very quickly. Here, however, we have a very "tough call." Alternately, we need to get more people involved and in a state or readiness to get and provide more/better photographs of the creature. With additional photographs we can probably tip the scale. Incidentally, Roger Patterson, Bob Gimlin and Bob Titmus are the only researchers I know of who went looking for the creature and found it. All other sightings per se are just plain chance encounters with everyday people.

I would venture to say that there have been numerous opportunities for good photographs had the observer been mentally ready for an encounter and equipped with a camera. So the "call to action" is really very simple. It is the same as that of the Boy Scouts, "Be Prepared."

Finally, I will mention that we believe a considerable amount of evidence, even good photographs, has not been brought to light. Reasons for this situation vary from plain disinterest to fear of publicity. I firmly encourage anyone with information to come forward, we need your help.

> "He who seeks long enough and hard enough will find the truth, whatever that truth may be."
>
> *Roger Patterson*

> "I firmly encourage anyone with information to come forward, we need your help."

BIBLIOGRAPHY

SUBJECT BOOKS, GENERAL BOOKS, MAGAZINE ARTICLES, ESSAYS, PAPERS AND WEBSITES

Alley, J. Robert, 2003, *Raincoast Sasquatch*, Hancock House Publishers, Surrey, B.C., Canada

Bayanov, Dmitri, 1996, *In the Footsteps of the Russian Snowman*, Crypto Logos Publishers, Moscow, Russia

Bayanov, Dmitri, 1997, *America's Bigfoot: Fact Not Fiction,* Crypto Logos Publishers, Moscow, Russia

Bayanov, Dmitri, 2001, *BIGFOOT: To Kill or to Film, The Problem of Proof,* Pyramid Publications, Burnaby, B.C., Canada

Beck, Fred, 1967, *I Fought the Apemen of Mt. St. Helens*, R.A. Beck, Kelso, Washington, U.S.A./Pyramid Publications, New Westminster, B.C., Canada

Bindernagel, John, A., 1998, *North America's Great Ape: The Sasquatch*, Beachcomber Books, Courtenay, B.C. Canada

Bourtsev, Igor, 2003, *Hominology in Russia – Personal Observations, Problems and Perspectives,* I. Bourtsev, Moscow, Russia (Paper)

Burns, John W., 1954, *My Search for B.C.'s Giant Indians*, Liberty magazine, December

Coleman, Loren, 1989, *Tom Slick and the Search for the Yeti*, Faber and Faber, Winchester, Massachusetts, U.S.A.

Crowe, Raymond, The Track Record (monthly publication), R. Crowe, Hillsboro, Oregon, U.S.A.

Donskoy, Dmitri D., 1971, *Qualitative Biomechanical Analysis of the Walk of the Creature in the Patterson Film,* (Paper)

Fahrenbach, W. H., 1997/98, *Sasquatch Size, Scaling and Statistics*, Cryptozoology (journal), Volume 13

Glickman, J., 1998, *Toward a Resolution of the Bigfoot Phenomenon*, North American Science Institute, Hood River, Oregon, U.S.A.

Green, John, 1969/80/94, *On the Track of the Sasquatch*, Cheam Publishing, Agassiz, B.C., Canada/Hancock House Publishers, Surrey, B.C. Canada

Green, John, 1973, *The Sasquatch File*, Cheam Publishing, Agassiz, B.C. Canada

Green, John, 1970, *The Year of the Sasquatch*, Cheam Publishing, Agassiz, B.C., Canada

Green, John, 1981, *Sasquatch, The Apes Among Us*, Hancock House Publishers, Surrey, B.C., Canada

Green, John, 1980, *Encounters with Bigfoot*, Hancock House Publishers, Surrey, B.C. Canada

Green, John, 2004, *The Best of Sasquatch/Bigfoot*, Hancock House Publishers, Surrey, B.C., Canada

Grieve, Donald W. (1971) *Report on the Film of a Proposed Sasquatch,* (Paper)

Halpin, Marjorie; Ames, Michael M.,1980, *Manlike Monsters on Trial, Early Records and Modern Evidence*, University of British Columbia Press, Vancouver, B.C., Canada

Hunter, Don and R. Dahinden, 1973, *Sasquatch/Bigfoot, The Search for North America's Incredible Creature*, McClelland and Stewart, Toronto, Ontario, Canada

Kane, Paul, 1925, *Wanderings of an Artist Among the Indians of North America, Th*e Radisson Society of Canada Ltd, Toronto, Ontario, Canada

Krantz, Grover S., 1992, *BIG FOOTPRINTS, A Scientific Inquiry into the Reality of Sasquatch*, Johnson Printing Co., Boulder, Colorado, U.S.A.

Krantz, Grover S., 1999, *Bigfoot/Sasquatch Evidence*. Hancock House Publishers, Surrey, B.C., Canada.

Long, Greg, 2004, *The Making of Bigfoot*, Prometheus Books, New York, New York, U.S.A.

Meldrum, D. Jeffrey, 2003, *Evaluation of Alleged Sasquatch Footprints and Their Inferred Functional Morphology*, (Paper/poster)

Meldrum, D. Jeffrey, 2003, *Dermatoglyphics in Casts of Alleged North American Ape Footprints* (Paper/poster)

Moneymaker, Matthew, The Bigfoot Field Researchers Organization (Website)

Murphy, Christopher; Cook, Joedy; Clappison, George, 1997, *Bigfoot in Ohio, Encounters with the Grassman*, Pyramid Publications, New Westminster, B.C. Canada

Murphy, Daniel, 1995, *Bigfoot: More than Meets the Eye*, Scott Stamp Monthly magazine, April (Cover story)

Napier, John, 1972, *BIGFOOT, Startling Evidence of Another Form of Life on Earth*, Berkley Publishing, New York, New York, U.S.A.

Perez, Daniel, *Bigfoot Times Newsletter* (monthly publication), Norwalk, California, U.S.A.

Powell, Thom, 2003, *The Locals, A Contemporary Investigation of the Bigfoot/Sasquatch Phenomenon*, Hancock House Publishers, Surrey, B.C. Canada

Roosevelt, Theodore, 1893, *The Wilderness Hunter – Outdoor Pastimes of an American Hunter*, G. P. Putnan's Sons, New York, New York, *U.S.A.*

Rense, Jeff, 2004, The Jeff Rense Radio Program – Interview with Greg Long, Robert Kiviat, Karl Korff and Robert Hieronimus, March 1

Sanderson, Ivan T., 1959, *The Strange Story of America's Abominable Snowman, True* magazine, December

Sanderson, Ivan T., 1969, *First Photos of 'Bigfoot,' California's Legendary Abominable Snowman*, Argosy magazine, February

Short, Bobbie, Bigfoot Encounters (Website)

Staff Writer, 1974, *The Search Goes on for Bigfoot, The Smithsonian*, Smithsonian Institution, Washington, D.C., January, Volume 4, No. 10

Steenburg, Thomas, 1990, *The Sasquatch in Alberta*, Western Publishers, Calgary,Alberta, Canada

Steenburg, Thomas, 2000, *In Search of Giants, Bigfoot Sasquatch Encounters*, Hancock House Publishers, Surrey, B.C. Canada

Steenburg, Thomas, 1993, *Sasquatch/Bigfoot, The Continuing Mystery*, Hancock House Publishers, Surrey, B.C. Canada

Thompson, David, *David Thompson: Narrative of his Explorations in Western America*, 1784-1812, Greenwood Press, Westport, Connecticut, U.S.A.

Wasson, Barbara, 1979, *Sasquatch Apparitions*, Barbara Wasson, Bend, Oregon, U.S.A.

REFERENCE BOOKS

Chronicle of Canada, 1990, Chronicle Publications, Montreal, Quebec, Canada

Washington Environmental Atlas, 1975, United States Army Corps, U.S. Government

NEWSPAPER ARTICLES

Colonist, Victoria, British Columbia, Canada, July 3, 1882, "What is it? A Strange Creature Captured Above Yale; A British Columbia Gorilla," Correspondence to the *Colonist*, by staff reporter

Humboldt Times, Humboldt, California, U.S.A., October 14, 1958, "Huge Footprints Hold Mystery of Friendly Bluff Creek Giant" by Andrew Genzoli

Memphis Enquirer, Memphis, Tennessee, U.S.A., May 9, 1851, "Wild Man of the Woods" by staff reporter

Province, Vancouver B.C. Canada, October 6, 1958, "New 'Sasquatch' Found – It's called Bigfoot" by staff reporter

San Jose News, San Jose, California c1959, "Huge Caveman Loose? 17-inch Footprint!" by Jeannette Befame

Times, London, England February 4, 2001, "Yeti Hair Defies DNA Analysis," by Mark Henderson

Times-Standard, Eureka, California, October 21, 1967, "Mrs. Bigfoot Is Filmed," by staff reporter

Washington Post, Washington, D.C., U.S.A., March 7, 2004, "Sasquatch Speaks: The Truth is Out" by Richard Leiby

Washington Star News, Washington State, U.S.A., July 1975, "Recognition at Last," by staff reporter

Yakima Herald – Republic, Yakima, Washington, January 30, 1999, "Bigfoot Unzipped, Man Claims It Was Him in a Suit!" by David Wasson

TELEVISION DOCUMENTARIES

The World's Greatest Hoaxes (1998), BBC

Sasquatch: Legend Meets Science (2003), Whitewolf Entertainment Inc

Photograph/Illustration - Sources/Copyrights

PG	DESCRIPTION	COPYRIGHT/CREDIT	ARTIFACT COLLECTION
0	COVER - BIGFOOT HEAD	P. TRAVERS	
8	P. TRAVERS	P. TRAVERS	
8	Y. LECLERC	Y. LECLERC	
10	STONE HEAD - ACTUAL PHOTO	UNIV. OF OREGON	U/O MUSEUM OF NAT HISTORY, OR
10	STONE HEAD - FIRST DRAWING (T)	P.TRAVERS	PEABODY MUSEUM, YALE UNIV., CT
10	STONE HEAD - SECOND DRAWING (L)	P. TRAVERS	AMERICAN MUS. OF NAT. HISTORY, N.Y.
11	PORTRAIT OF A SASQUATCH	Y. LECLERC	
12	STONE FOOT - TWO VIEWS	J. GREEN	VAN. MUS. BC CAT.# QAD92
13	PETROGLYPH - BELLA COOLA	C. MURPHY	
13	PETROGLYPH - PAINTED ROCK	K. MOSCOWITZ	
14	PETROGLYPH - WISTLING MOUTH	R. MORGAN	
14	PETROGLYPH - FOOTPRINTS	R. MORGAN	
15	PETROGLYPH - SATISFIED SASQUATCH	R. MORGAN	
15	PETROGLYPH - POWER STONE	R. MORGAN	
15	PETROGLYPH - GIANTESS/HUMAN	R. MORGAN	
15	PETROGLYPH - STONE PLUGS	R. MORGAN	
16	MAP ILLUSTRATION	Y. LECLERC	
16	SASQUATCH DRAWING	P. TRAVERS	
17	MASK - TSIMSHIAN, FIRST DRAWING (LEFT)	P. TRAVERS	PEABODY MUSEUM, HARVARD UNIV., MA
17	MASK - NISHGA SECOND DRAWING (CENTER)	P. TRAVERS	ROYAL ONTARIO MUSEUM
17	MASK - NISHGA THIRD DRAWING (RIGHT)	P. TRAVERS	ROYAL BC PROV. MUSEUM CAT.# 9717
17	MASK KWAKUITL	J. GREEN	ROYAL BC PROV. MUSEUM
17	KWAKIUTL HERALDIC POLE	JOHN GREEN	ROYAL BC PROV. MUSEUM
17	CHEHALIS MASK - FULL FACE	C. L. MURPHY	VAN., MUS, BC CAT.# AA69
17	CHEHALIS MASK - PROFILE	C. L. MURPHY	VAN., MUS, BC CAT.# AA69
22	CONTEMPORY KWAKUITL DESIGN	D. HANCOCK	
22	CONTEMPORY SASQUATCH DOLL	D. HANCOCK	
22	TRANSFORMATION MASK - TOP	D. HANCOCK	
22	TRANSFORMATION MASK - LOWER	D. HANCOCK	
22	D'SONOQUA POLE	D. HANCOCK	
23	TSUNGANI (DANCING FIGURE)	D. HANCOCK	
18	DELAWARE CARVING	J. COOK	
18	KWAKIUTL MASK	C. L. MURPHY	J. ROBERT ALLEY, AK
18	HAIDA MASK	C. L. MURPHY	J. ROBERT ALLEY, AK
18	TREE CARVING	R. MORGAN	
19	PICTOGRAPHS - FULL PHOTO	K. MOSCOWITZ	
19	PICTOGRAPHS - FULL PHOTO DIAGRAM	K. MOSCOWITZ	
20	PICTOGRAPH PHOTO DETAIL MOTHER	K. MOSCOWITZ	
20	PICTOGRAPH PHOTO DIAGRAM MOTHER	K. MOSCOWITZ	
20	PICTOGRAPH PHOTO DETAIL CHILD	K. MOSCOWITZ	
20	PICTOGRAPH PHOTO DIAGRAM CHILD	K. MOSCOWITZ	
22	MAP DIAGRAM	K. MOSCOWITZ	
21	PICTOGRAPH PHOTO DETAIL - CATERPILLER	K. MOSCOWITZ	
21	PICTOGRAPH PHOTO DETAIL - COYOTE	K. MOSCOWITZ	
21	PICTOGRAPH PHOTO DETAIL - PEOPLE	K. MOSCOWITZ	
4	POSTAGE STAMP - THOMPSON	CANADA POST	
24	PAUL KANE	PUBLIC DOMAIN	
25	THEODORE ROOSEVELT	PUBLIC DOMAIN	
26	YALE, B.C., CIRCA 1882	PUBLIC DOMAIN	
28	FRED BECK	JOHN GREEN	
29	BOOK COVER - APEMAN	E. DAVENPORT/R. BECK	
29	APE CANYON	JOHN GREEN	
30	OSTMAN SCRIBBLER	C. MURPHY	JOHN GREEN, BC
30	DAHINDEN AND OSTMAN	E. DAHINDEN	
28	OSTMAN SCRIBBLER	C. MURPHY	JOHN GREEN, BC
31	CHEHALIS LOGO	CHEHALIS PEOPLE	
34	R. DAHINDEN & C. MURPHY	C. MURPHY	
34	CHEHALIS RIVER	C. MURPHY	
34	CHEHALIS ADMIN. BUILDING	C. MURPHY	
34	BRAD TOMBE	B. TOMBE	
34	ALEXANDER PAUL & D. MURPHY	C. MURPHY	
34	BRAD TOMBE LETTER	B. TOMBE	
35	TRACING OF CAST	C. MURPHY	JOHN GREEN, BC
35	RUBY CREEK AREA SCENE	C. MURPHY	
35	DAHINDEN STEPPING OVER FENCE	J. GREEN	
35	CHAPMAN HOUSE - DISTANT VIEW	J. GREEN	
35	CHAPMAN HOUSE - CLOSE-UP VIEW	J. GREEN	
36	DRAWING BY ROE'S DAUGHTER	J. GREEN	
37	JERRY CREW - TOP PHOTO	VANCOUVER PROVINCE	
37	JERRY CREW - LOWER PHOTO	HUMBOLDT TIMES	
38	WILLOW CREEK, 1960	WILLOW CREEK MUSEUM	
39	PACIFIC NORTHWEST EXPEDITION	J. GREEN	
40	FRAME 364 - P/G FILM	E. DAHINDEN	
40	BOOK COVER - R. PATTERSON'S BOOK	E. DAHINDEN	
40	COVER OF TRUE MAGAZINE	MAGAZINE (DEFUNCT)	TOM COUSINO COLLECTION, CA
40	SCULPTURE BY ROGER PATTERSON	C. MURPHY	PATRICIA PATTERSON, WA
41	MAP ILLUSTRATION - CALIFORNIA	Y. LECLERC	
42	MAP OF THE FILM SITE	R. TITMUS	
43	PATTERSON WITH CASTS (1967)	E. DAHINDEN	
43	TIMES-STANDARD HEADING (1967)	TIMES-STANDARD	
45	FRAME 352 - P/G FILM (FULL FRAME)	E. DAHINDEN	
46	PATTERSON MAKING A CAST	E. DAHINDEN	
46	PATTERSON HOLDING CASTS (1970)	E. DAHINDEN	
46	STRIP OF 10MM FILM	C. MURPHY	
47	GIMLIN HOLDING CASTS	E. DAHINDEN	
47	GIMLIN AND PATTERSON	E. DAHINDEN	
47	GIMLIN IN 2003	C. MURPHY	
48	FOOTPRINT - LEFT SET FIRST	E. DAHINDED	
48	FOOTPRINT - LEFT SET SECOND	R. LYLE LAVERTY	
48	FOOTPRINT - LEFT SET THIRD	R. LYLE LAVERTY	
48	FOOTPRINT - LEFT SET FOURTH	R. LYLE LAVERTY	
48	FILM SITE CASTS - W/HUMAN CAST	E. DAHINDEN	E. DAHINDEN BC
48	FILM SITE CASTS - W/HUMAN FOOT	E. DAHINDEN	E. DAHINDEN BC
50	ARGOSY MAGAZINE COVER	MAGAZINE (DEFUNCT)	
51	FRAME 61 - P/G FILM	E. DAHINDEN	
51	FRAME 72 - P/G FILM	E. DAHINDEN	
51	FRAME 307 - P/G FILM	E. DAHINDEN	
51	FRAME 310 - P/G FILM	E. DAHINDEN	
51	FRAME 323 - P/G FILM	E. DAHINDEN	
51	FRAME 332 - P/G FILM	E. DAHINDEN	
51	FRAME 339 - P/G FILM	E. DAHINDEN	
51	FRAME 343 - P/G FILM	E. DAHINDEN	
51	FRAME 350 - P/G FILM	E. DAHINDEN	
51	FRAME 352 - P/G FILM	E. DAHINDEN	
51	FRAME 362 - P/G FILM	E. DAHINDEN	
51	FRAME 364 - P/G FILM	E. DAHINDEN	
52	FRAME 61-P/G FILM - ENLARGED	E. DAHINDEN	
52	FRAME 72-P/G FILM - ENLARGED	E. DAHINDEN	
52	FRAME 307-P/G FILM - ENLARGED	E. DAHINDEN	
52	FRAME 310-P/G FILM - ENLARGED	E. DAHINDEN	
53	FRAME 323-P/G FILM - ENLARGED	E. DAHINDEN	
53	FRAME 332-P/G FILM - ENLARGED	E. DAHINDEN	
53	FRAME 339-P/G FILM - ENLARGED	E. DAHINDEN	
53	FRAME 342-P/G FILM - ENLARGED	E. DAHINDEN	
54	FRAME 350-P/G FILM - ENLARGED	E. DAHINDEN	
54	FRAME 352-P/G FILM - ENLARGED	E. DAHINDEN	
54	FRAME 362-P/G FILM - ENLARGED	E. DAHINDEN	
54	FRAME 364 -P/G FILM - ENLARGED	E. DAHINDEN	
55	FRAME 61 -P/G FILM - FULL FRAME	E. DAHINDEN	
55	FRAME 72 -P/G FILM - FULL FRAME	E. DAHINDEN	
55	FRAME 310 -P/G FILM - FULL FRAME	E. DAHINDEN	
55	FRAME 323 -P/G FILM - FULL FRAME	E. DAHINDEN	
55	FRAME 339 -P/G FILM - FULL FRAME	E. DAHINDEN	
55	FRAME 343 -P/G FILM - FULL FRAME	E. DAHINDEN	
55	FRAME 350 -P/G FILM - FULL FRAME	E. DAHINDEN	
55	FRAME 352 -P/G FILM - FULL FRAME	E. DAHINDEN	
55	FRAME 362 -P/G FILM - FULL FRAME	E. DAHINDEN	
56	FILM SITE AERIAL VIEW	E. DAHINDEN	
57	KODAK CAMERA AD	EASTMAN KODAK CO	
57	CAMERA WITH RULER	C. MURPHY	D. PEREZ, CA
57	CAMERA IN OPEN POSITION	C. MURPHY	D. PEREZ, CA
58	FILM SITE MODEL - FULL VIEW	C. MURPHY	C. MURPHY. BC
58	FRAME 352 WITH KEY LINES	E. DAHINDEN	
58	FILM SITE DIAGRAM	E. DAHINDEN	
59	FRAME 352 - P/G FILM WITH ARROW	E. DAHINDEN	
60	FRAME 352- P/G FILM WITH ARROW RPT	E. DAHINDEN	
60	FRAME 352-P/G FILM W/FILM SITE MODEL	E. DAHINDEN/C.MURPHY	
61	FILM SITE MODEL - EXTERNAL POINTS	C. MURPHY	
62	FILM SITE MODEL - INTERNAL POINTS	C. MURPHY	
63	FILM SITE MODEL - LEFT VIEW	C. MURPHY	
63	FRAME 364 - P/G FILM HEAD DETAIL	E. DAHINDEN	
63	FILM SITE MODEL - RIGHT VIEW	C. MURPHY	
63	FILM SITE MODEL - BACK VIEW	C. MURPHY	
63	FRAME 384 - P/G FILM HEAD DETAIL	E. DAHINDEN	
64	J. GREEN & C. MURPHY	C. MURPHY	
64	R. GIMLIN, C. MURPHY & FILM SITE MODEL	C. MURPHY	
65	CREATURE MEASUREMENTS	J. GREEN/Y. LECLERC	
66	FOOTPRINTS IN A SERIES (FIRST PHOTO)	E. DAHINDEN/Y. LECLERC	
66	FOOTPRINTS IN A SERIES (SECOND PHOTO)	E. DAHINDEN/Y. LECLERC	
66	R. PATTERSON MAKING CAST WITH LINE	E. DAHINDEN	
67	FRAME 323 P/G FILM W/PRINTS MARKED	E. DAHINDEN/C. MURPHY	
67	FRAME 291 P/G FILM (FOOT ON FRAGMENT)	E. DAHINDEN	
68	FRAME 352 WITH ARROW	E. DAHINDEN	
68	WOOD FRAGMENT - FILM POSITION	C. MURPHY	E. DAHINDEN BC
68	WOOD FRAGMENT - WITH RULER	C. MURPHY	E. DAHINDEN BC
70	HEAD - FIRST PANE - UPPER LEFT	E. DAHINDEN/C.MURPHY	
70	HEAD - SECOND PANE - UP RIGHT	C. MURPHY	
70	HEAD - THIRD PANE - LOWER LEFT	C. MURPHY	
70	HEAD - FOURTH PANE - LOWER RIT	C. MURPHY	
71	HEAD - FRAME 339 - LEFT	E. DAHINDEN/C.MURPHY	
71	HEAD - FRAME 339 - RIGHT	C. MURPHY/Y. LECLERC	
71	HEAD - FRAME 350 - LEFT	E. DAHINDEN/C. MURPHY	
71	HEAD - FRAME 350 - RIGHT	Y. LECLERC	
71	P. BIRNAM & C. MURPHY W/HEADS	C. MURPHY	P. BIRMHAM BC
71	SCULPTURED HEAD - ENLARGEMENT	C. MURPHY	P. BIRMHAM BC
72	DMITRI BAYANOV	D. PEREZ	

Page	Description	Credit	Additional
72	I. BOURTSEV	D. PEREZ	
74	DR. D. DONSKOY	PUBLIC DOMAIN	
76	DR. D.W. GRIEVE	E. DAHINDEN	
80	J. GLICKMAN - PORTRAIT	J. SEMLOR, HOOD RIVER NEWS	
81	J. GLICKMAN AT COMPUTER	J. SEMLOR, HOOD RIVER NEWS	
81	SASQUATCH HEAD WITH RIFLE SCOPE	C. MURPHY/ Y. LECLERC	
81	SASQUATCH HEIGHT ILLUSTRATION	E. DAHINDEN/ Y.LECLERC	
82	DR. G. KRANTZ	J. GREEN	
90	DIGITIZED SASQUATCH (3 IMAGES)	D. HAJICEK	
90	DATA POINTS - TOES & FEET	D. HAJICEK	
90	REUBEN STEINDORF	D. HAJICEK	
91	BONES - TOP LEFT	D. HAJICEK	
91	D. HAJICEK & MARIO BENASSI	D. HAJICEK	
91	DEREK PRIOR	D. HAJICEK	
91	PELVIS/LEG BONES	D. HAJICEK	
91	SKELETON ON VIRTUAL EXERCISE WHEEL	D. HAJICEK	
91	SKELETON, LOWER SET PHOTOS - LEFT	D. HAJICEK	
91	SKELETON, LOWER SET PHOTOS - CENTER	D. HAJICEK	
91	SKELETON, LOWER SET PHOTOS - RIGHT	D. HAJICEK	
92	BONES	D. HAJICEK	
92	SCENE - FILM SITE, LEFT	D. HAJICEK	
92	SCENE - FILM SITE, RIGHT	D. HAJICEK	
93	BONES - PELVIS/LEGS, LEFT	D. HAJICEK	
93	BONES - PELVIS/LEGS, CENTER	D. HAJICEK	
93	BONES - PELVIS/LEGS, RIGHT	D. HAJICEK	
93	FRAME 350, P/G FILM, HEAD DETAIL	E. DAHINDEN	
93	FRAME 350, HEAD ENHANCEMENT	P. TRAVERS	
93	FULL DIGITAL FACE	D. HAJICEK	
93	DIGITAL FACE MERGE	D. HAJICEK	
93	FINAL DIGITAL FACE	D. HAJICEK	
93	SCENE - FILM SITE, - LEFT	D. HAJICEK	
93	SCENE - FILM SITE, CENTER	D. HAJICEK	
93	SCENE - FILM SITE, RIGHT	D. HAJICEK	
94	MULTIPLE VIEWS - SIX IMAGES	D. HAJICEK	
95	DIGITAL FOOTPRINT	D. HAJICEK	
95	DIGITAL MAP	D. HAJICEK	
95	VIRTUAL 3D PLASTER CAST	D. HAJICEK	
95	HANDHELD SCANNER	D. HAJICEK	
95	PATH - MEMORIAL DAY CREATURE	D. HAJICEK	
95	LIDAR LASER SCANNER	D. HAJICEK	
95	300 MILLION DATA POINTS	D. HAJICEK	
96	I. BOURTSEV WITH SCULPTURE	I. BOURTSEV	
96	FRAME 352 - P/G FILM	E. DAHINDEN	
96	SCULPTURE - BY I. BOURTSEV	E. DAHINDEN	E. DAHINDEN, BC
97	SASQUATCH PAINTING	R. BATEMAN	
98	HEAD, PROFILE	Y. LECLERC	
98	HEAD, FULL FACE CONSTRUCTION	Y. LECLERC	
98	MOTHER AND CHILD	Y. LERCLERC	
99	HEAD W/IDENTIFIED CHARACTERISTICS	E. DAHINDEN/C.MURPHY	
99	FRAME 352 - P/G FILM WITH CIRCLES	E. DAHINDEN/C. MURPHY	
99	HEAD MODEL, FRONT VIEW	C. MURPHY	C. MURPHY, BC
99	HEAD MODEL, SEMI-SIDE VIEW	C. MURPHY	C. MURPHY, BC
99	HEAD MODEL, NO HAIR (CONST. PHOTO)	C. MURPHY	
100	HEAD - YOUNG SASQUATCH, LEFT PANE	C. MURPHY	POSSESSION OF ARTIST
100	HEAD - YOUNG SASQUATCH, RIGHT PANE	C. MURPHY/Y. LECLERC	
100	SASQUATCH ARM, PALM DOWN	C. MURPHY	WILD ENT/ WILDERNESS PROD., BC
100	SASQUATCH ARM, PALM UP	C. MURPHY	
100	C. MURPHY WITH SASQUATCH ARM	C. MURPHY	
101	STAMPS - LEGENDARY CREATURES	CANADA POST	
101	STAMP - SINGLE - SASQUATCH	CANADA POST	
101	COVER - STAMP MONTHLY MAGAZINE	SCOTT PUBLISHING CO	
102	HOW FOOTPRINT CASTS ARE MADE	C. MURPHY	C. MURPHY, BC
103	SASQUATCH FOOT WITH HUMAN FOOT	C. MURPHY	C. MURPHY, BC
103	SASQUATCH FOOT FACING LEFT	C. MURPHY	C. MURPHY, BC
103	SASQUATCH FOOT - FRONT VIEW	C. MURPHY	C. MURPHY, BC
104	CAST MAKING BOX - VERTICLE VIEW	C. MURPHY	C. MURPHY, BC
104	CAST MAKING BOX - HORZORTAL VIEW	C. MURPHY	C. MURPHY, BC
104	HUMAN FOOT CAST	C. MURPHY	C. MURPHY. BC
104	FOOT TRANSPARENCY ON CAST	C. MURPHY	
104	SASQUATCH FOOT ILLUSTRATION	J. MELDRUM	
105	CAST #1, J. CREW, 1958, BLUF CREEK	C. MURPHY	C. MURPHY, BC (SEE NOTE 1)
105	CAST #2, J. GREEN, 1967, BLUFF CREEK	C. MURPHY	C. MURPHY, BC (SEE NOTE 1)
105	CAST #3, J. GREEN, 1967, BLUE CRK MTN	C. MURPHY	C. MURPHY, BC (SEE NOTE 1)
105	CAST #4, POSSIBLE BLUFF CREEK, 1960s	C. MURPHY	C. MURPHY, BC
105	CAST #5, DR. J. BINDERDAGEL, 1988, SPP - VAN. IS.	C. MURPHY	VANCOUVER MUSEUM, BC (SEE NOTE 2)
105	CAST #6, A.D. HERYFORD, 1982, OLMPIC PENN	C. MURPHY	R. NOLL, WA
106	CAST #7, J. COOK, 2003 SHAWNEE ST. PK, OHIO	C. MURPHY	J. COOK, OHIO
106	CAST #8, T. STEENBURG, 1986, CHILLIWACK	C. MURPHY	C. MURPHY, BC (SEE NOTE 3)
106	CAST #9, R. PATTERSON, 1964, BLUFF CREEK	C. MURPHY	C. MURPHY, BC (SEE NOTE 3)
106	CASTS #10 (SET), R.TITMUS, 1958, BLUFF CREEK	C. MURPHY	C. MURPHY, BC (SEE NOTE 4)
106	CAST #11A, R.TITMUS, 1963, HYAMPOM	C. MURPHY	WILLOW CREEK CHINA FLAT MUSEUM, CA
106	CAST #11B, R.TITMUS, 1963, HYAMPOM	C. MURPHY	WILLOW CREEK CHINA FLAT MUSEUM, CA
106	CAST #11C, R.TITMUS, 1963, HYAMPOM	C. MURPHY	C. MURPHY, BC (SEE NOTE 1)
106	CAS #11D, R.TITMUS, 1963, HYAMPOM	C. MURPHY	WILLOW CREEK CHINA FLAT MUSEUM, CA
106	CAST #11E, R.TITMUS, 1963, HYAMPOM	C. MURPHY	WILLOW CREEK - CHINA FLAT MUSEUM, CA
106	CASTS #12 (SET), R.TITMUS, 1976 SKEENA RIVER	C. MURPHY	C. MURPHY, BC (SEE NOTE 1)
107	CASTS #13 (SET), R. PATTERSON, 1967, BLUFF CREEK	C. MURPHY	E. DAHINDEN, BC
107	CAST #14A, R.TITMUS, 1967, BLUFF CREEK	C. MURPHY	C. MURPHY, BC (SEE NOTE 1)
107	CAST #14B, RTITMUS, 1967, BLUFF CREEK	C. MURPHY	C. MURPHY ,BC (SEE NOTE 1)
107	CAST #14C, R.TITMUS, 1967, BLUFF CREEK	C. MURPHY	C. MURPHY,BC (SEE NOTE 1)
107	CAST #14D, R.TITMUS, 1967, BLUFF CREEK	C. MURPHY	C. MURPHY,BC (SEE NOTE 1)
107	CAST #14E, R.TITMUS, 1967, BLUFF CREEK	C. MURPHY	C. MURPHY, BC (SEE NOTE 1)
107	CASTS #15 (SET), R. DAHINDEN, 1959, BOSSBURG	C. MURPHY	E. DAHINDEN, BC
107	CASTS #16 (SET), #15 WITH BONES	C. MURPHY	C. MURPHY ,BC (SEE NOTE 5)
107	CAST #17, P. FREEMAN. 1982, WALLA WALLA	C. MURPHY	C. MURPHY, BC (SEE NOTE 6)
108	CAST, J.CREW, 1958 - TITMUS (CLEANED CAST)	C. MURPHY	VANCOUVER MUSEUM, BC
108	CAST, J. BINDERNAGEL, (SAME AS #5) WITH BONES	Y. LECLERC	
108	FEET ILLUSTRATION	Y. LECLERC	
109	CASTS & WALLACE WOODEN FOOT	R. NOLL	R. NOLL, WA
110	PRINT, 13-INCHES, BLUE CREEK MTN., 1967	E. DAHINDEN	
110	PRINT, 15-INCHES, BLUE CREEK MTN., 1967	J. GREEN	
110	PRINT, 15 AND 13-INCHES, BLUE CREEK MTN, 1967	J. GREEN	
111	D. ABBOTT WITH PRINT	J. GREEN	
111	GREEN MEASURING PRINTS	J. GREEN	
111	PRINT WITH RULER	J. GREEN	
111	SASQUATCH PRINT WITH BOOT PRINT	J. GREEN	
112	R. TITMUS WITH CASTS	J. GREEN	
112	PRINTS - ELLESBERG, WA	J. GREEN	
112	CAST WITH HUMAN FOOT (BOOT)	J. GREEN	
112	R. TITMUS & S. MCCOY WITH CASTS	J. GREEN	
113	PRINTS, NOOKSACK ESTUARY	J. GREEN	
113	CAST SECTION WITH DERMAL RIDGES	MRS.G. KRANTZ	
114	PRINT WITH TAPE MEASURE	J. GREEN	
114	J. GREEN WITH CASTS	J. GREEN	
114	CAST, ABBOTT HILL	C. MURPHY	R. NOLL, WA
114	PRINTS, BLUFF CREEK, SPRINKLED WHITE	J. GREEN	
115	PRINT, 13-INCH WITH RULER	J. GREEN	
115	R. TITMUS WITH HYAMPOM CASTS ON WALK	J. GREEN	
115	R. TITMUS WITH JERRY CREW CAST	J. GREEN	
115	PRINT, HYAMPOM IN WET GROUND	J. GREEN	
115	PRINT, HYAMPOM, WITH YARD STICK	J. GREEN	
115	BERRYMAN, TITMUS AND MCCOY WITH CASTS	J. GREEN	
116	J. GREEN HOLDING CAST	J. GREEN	
116	PRINT IN SNOW - RULER, OHIO	ANONYMOUS/J. COOK	
116	PRINT, RULER & HUMAN FOOT	J. GREEN	
117	DR. G. KRANTZ & PRINT IN SNOW	J. GREEN	
117	PRINT, CRIPPLE FOOT, IN SNOW - CLOSE-UP	J. GREEN	
117	PRINT, CRUPPLE FOOT IN SOIL	J. GREEN	
117	R. DAHINDEN w/ CASTS	J. GREEN	
117	J. SUSEMIEHL w/ CASTS	J. GREEN	
117	N. DAVIS, WIFE & J. RHODES VIEWING PRINT	J. GREEN	
118	PRINT, OFFIELD MOUNTAIN - FULL VIEW	J. GREEN	
118	PRINT, OFFIELD MOUNTAIN - TOES ONLY	J. GREEN	
118	PRINT WITH BUCKET, SKEENA RIVER SLOUGH	R. TITMUS	
118	R. TITMUS WRITE-UP ON PRINTS	R. TITMUS	
118	STUMPS/ROOTS	R. TITMUS	
118	R. TITMUS HOLDING CASTS	R. TITMUS	
118	CAST DETAIL FROM R. TITMUS HOLDING CASTS	R. TITMUS	
119	PRINT W/ RULER/ BOOT PRINT, ONION MOUNTAIN	J. GREEN	
119	PRINT WITH RULER, BUNCOMBE HOLLOW	R. MORGAN	
119	R. MORGAN MEASURING PRINTS	R. MORGAN	
120	DR. R. MAURICE TRIPP WITH CAST	SAN JOSE NEWS (DEFUNCT)	
120	R. MORGAN AND DR. G. KRANTZ	R. MORGAN	
120	SOIL COMPACTION ILLUSTRATION	MRS. G. KRANTZ	
120	R. MORGAN HOLDING MEASURING STICK	R. MORGAN	
120	R. MORGAN DEMONSTRATING PACE	R. MORGAN	
121	PRINTS ON BLUE CREEK MOUNTAIN	J. GREEN	
121	HUMAN PRINTS ON BEACH	C. MURPHY	

Page	Description	Credit 1	Credit 2
21	PRINTS, ESTACADA, OREGON	J. GREEN	
21	PRINTS, POWDERMOUNTAIN, BC	J. GREEN	
21	PRINTS, DELOX MARSH, WISCONSIN	J. GREEN	
22	SASQUATCH FOOT & BEAR FOOT	E. DAHINDEN/ C. MURPHY	
22	DOUBLE TRACKED BEAR PRINT CAST & SASQ. CAST	C. MURPHY	C. MURPHY, BC (SEE NOTE 1)
22	DOUBLE TRACKED BEAR PRINT CAST ON GRID	G. KRANTZ	
22	PRINTS - HUMAN, SASQUATCH & BEAR	J. GREEN	
23	THE USUAL SUSPECTS POSTER	P. SMITH	
24	DR. H. FAHRENBACH	H. FAHRENBACH	
24	FOOT LENGTH GRAPH	H. FAHRENBACH	
25	FOOT WIDTH GRAPH	H. FAHRENBACH	
25	HEEL WIDTH GRAPH	H. FAHRENBACH	
25	FOOT WIDTH INDEX GRAPH	H. FAHRENBACH	
26	STEP LENGTH GRAPH	H. FAHRENBACH	
26	GROWTH GRAPH	H. FAHRENBACH	
26	GAIT GRAPH	H. FAHRENBACH	
27	FOOT TO HEIGHT RELATIONSHIP GRAPH	H. FAHRENBACH	
27	HEIGHT GRAPH	H. FAHRENBACH	
27	WEIGHT GRAPH	H. FAHRENBACH	
28	DR. H. FAHRENBACH WITH MICROPHONE	C. MURPHY	
28	D. MURPHY & Dr. H. FAHRENBACH	C. MURPHY	
28	GREEN, GIMLIN, CHILCUTT, Dr. FAHRENBACH	C. MURPHY	
29	DR. J. MELDRUM	C. MURPHY	
30	PRINTS, WALLA, WALLA - 12 IMAGES	J. MELDRUM	
31	PRINTS (2) WITH FOOT ILLUSTRATION	J. MELDRUM	
32	FRAME 61, PATTERSON/GIMLIN FILM	E. DAHINDEN	
32	PRINT WITH TWIG, BLUFF CREEK	R. L. LAVERTY	
32	SASQUATCH AND HUMAN FOOT ILLUSTRATION	J. MELDRUM	
33	PRINTS (3) WITH FOOT ILLUSTRATION	J. MELDRUM	
33	PRINTS, SATSOP RIVER - LEFT	D. HERYFORD	
33	HALF PRINT WITH RULER	H. FAHRENBACH	
33	FOOT ILLUSTRATION	J. MELDRUM	
34	CASTS, "CRIPPLEFOOT" WITH RULER	E. DAHINDEN	E. DAHINDEN BC
34	"CRIPPLEFOOT" DCNE ILLUSTRATION	J. MELDRUM	
35	PRINT WITH RULER, UPPER LEFT	E. DAHINDEN	
35	CAST WITH RULER, UPPER CENTER	J. MELDRUM	
35	CAST WITH RULER, UPPER RIGHT	J. MELDRUM	
35	CAST DETAIL	J. MELDRUM	
35	PRINT - SPLAYED TOE	J. MELDRUM	
35	TOE LENGTH CHART	J. MELDRUM	
36	FRAME 72, PATTERSON/GIMLIN FILM	E. DAHINDEN	
37	CAST, 13-INCH WITH DERMAL RIDGES	J. MELDRUM	
37	PRINT WITH RULER	J. MELDRUM	
38	TWO CASTS, HYAMPOM, WITH RULER	J. MELDRUM	
38	CAST DETAIL - DERMAL RIDGES	J. MELDRUM	
38	CAST DETAIL, LEFT, BLANCHARD, IDAHO	J. MELDRUM	
38	CAST DETAIL, RIGHT, BLANCHARD, IDAHO	J. MELDRUM	
39	CASTS (2) - SPLAYED TOES	J. MELDRUM	
39	HALF CAST	J. MELDRUM	
39	DRAWING OF CASTS	J. MELDRUM	
39	CAST DETAIL (DERMAL RIDGES)	J. MELDRUM	
40	FULL CAST (RIGHT PHOTO)	J. MELDRUM	
40	DRAWING OF FOOT	J. MELDRUM	
40	DERMAL RIDGES - LOWER RIGHT PHOTO	J. MELDRUM	
42	DERMAL RIDGE PATTERN EXAMPLES	R. NOLL/J. CHILCUTT/ C. MURPHY	
43	SASQ. KNUCKLES CAST AND HUMAN KNUCKLES	C. MURPHY	H. FAHRENBACH, OR
43	SASQUATCH HAND CAST AND HUMAN HAND	C. MURPHY	H. FAHRENBACH, OR
143	SASQ. HAND CAST/SASQ. HAND ILLUS./HUMAN HAND	Y. LECLERC/C. MURPHY	H. FAHRENBACH, OR
143	SASQ. KNUCKLE CASTST; SASQ ILL.; HUMAN HAND.	Y. LECLERC/C. MURPHY	H. FAHRENBACH, OR
144	HAND CAST (BOB TITMUS)	C. MURPHY	WILLOW CREEK - CHINA FLAT MUSEUM, CA
144	HUMAN HAND CAST	C. MURPHY	C. MURPHY, BC
144	SASQUATCH HAND CAST	C. MURPHY	H. FAHRENBACH, OR
144	SASQUATCH HAND PRINT AND HUMAN HAND	J. GREEN	
144	HAND PRINT IN MUD, OHIO	J. COOK	
144	GORILL/HUMAN/SASQUATCH HANDS	C.S. LINDLEY	C.S. LINDLEY, CA
145	SKOOKUM CAST & DR. JEFF MELDRUM	J. MELDRUM	R. NOLL EDMONDS, WA
145	KRANTZ/GREEN/ BINDERNAGEL & SKOOKUM CST.	R. NOLL	
145	RECLINED/REACHING SASQUATCH	P. TRAVERS	
145	SKOOKUM CAST W/ IDENTIFIED PARTS/MELDRUM	J. MELDRUM	
146/7	SKOOKUM CAST/SASQUATCH - LARGE PRINT	R. NOLL	
148	DIAGRAM #1 SHOWING EARTH/MUD	P. TRAVERS	
148	DIAGRAM #2 SHOWING MAN AND FRUIT	P. TRAVERS	
148	DIAGRAM #3 SHOWING SASQ. AT EARTH/MUD	P. TRAVERS	
148	DIAGRAM #4 SHOWING SASQUATCH REACHING	P. TRAVERS	
149	HEEL FROM THE SKOOKUM CAST	C. MURPHY	R. NOLL, EDMONDS, WA
149	HUMAN FOOT IN SAND	C. MURPHY	
149	HEE; DETAIL FROM SKOOKUM CAST	R. NOLL	
150	SKOOKUM HEEL & HUMAN HEEL	R. NOLL	
150	LOCATION OF SKOOKUM CAST DIAGRAM	R. NOLL/P. TRAVERS	
151	DRS. MELDRUM, SARMIENTO, SWINDLER	D. HAJICEK	
151	DR. DARIS SWINDLER HOLDING JAW BONES	J. GREEN	
151	R. NOLL WITH CASTS	R. NOLL	
153	PAGE FROM WASHINGTON ENV. ATLAS	PUBLIC DOMAIN	
154	J. COOK	J. COOK	
154	G. CLAPPISON	J. COOK	
155	HAIR SAMPLE - DEER	H. FAHRENBACH	
155	HAIR SAMPLE - CHIMPANZEE	H. FAHRENBACH	H. FAHRENBACH
155	HAIR SAMPLE - HUMAN	H. FAHRENBACH	H. FAHRENBACH
155	HAIR SAMPLE - SASQUATCH, CA	H. FAHRENBACH	H. FAHRENBACH
155	HAIR SAMPLE - SASQUATCH, WA #1	H. FAHRENBACH	H. FAHRENBACH
155	HAIR SAMPLE - SASQUATCH, WA #2	H. FAHRENBACH	H. FAHRENBACH
156	CLOSE-UP - KIAWOCK LAKE NEST	E. MUENCH	
156	DISTANT VIEW - KIAWOCK LAKE NEST	E. MUENCH	
157	LARGE CEDAR TREE	E. MUENCH	
157	PARALLEL 8-IN. SPAN MARKS	E. MUENCH	
158	NEST IN TREES	J. COOK	
159	HOLLOW AND G. CLAPPISON STANDING	J. COOK	
159	HOLLOW AND G. CLAPPISON PEERING IN SIDE	J. COOK	
159	HOLLOW AND G. CLAPPISON INSIDE	J. COOK	
159	J. COOK	C. MURPHY	
159	BOOK COVER - BIGFOOT IN OHIO	C. MURPHY	
160	R. MOREHEAD	R. MOREHEAD	
161	JIM GREEN IN ROCKS	J. GREEN	
161	G. THOMAS	R. NOLL	
161	DISTANT VIEW OF ROCKS	R. NOLL	
162	ELEVATED VIEW OF ROCKS	R. NOLL	
162	ROCK PILES	R. NOLL	
162	BOOK COVER - N.A.'s. GREAT APE - SASQUATCH	J. BINDERNAGEL	
163	DR. J. BINDERNAGEL	J. BINDERNAGEL	
163	WOLVERINE	D. HANCOCK	
164	SASQUATCH PROFILE	Y. LECLERC	
164	FOUR SKULLS (TOP)	Y. LECLERC	
164	FOUR SUPERIMPOSED SKULLS	Y. LECLERC	
165	COMPOSITE SKULL (3 ILLUSTRATIONS)	Y. LECLERC	
165	ACTUAL FILM FRAME NO. 343 DETAIL	E. DAHINDEN/C. MURPHY	
165	DEFINED HEAD	E. DAHINDEN/Y. LECLERC	
165	SUPERIMPOSED SKULL (2 ILLUSTRATIONS)	E. DAHINDEN/Y. LECLERC	
166	CREATURE HEIGHT CHART	E. DAHINDEN/Y. LECLERC	
167	MAP OF SASQ. INCIDENTS - BRITISH COLUMBIA	J. GREEN	
168	MAP OF SASQ. INCIDENTS - WASHINGTON	J. GREEN	
169	MAP OF SASQ. INCIDENTS - CALIFORNIA	J. GREEN	
170	MAP OS SQSQ. INCIDENTS - MONTANA	J. GREEN	
171	MAP OF SASQ. INCIDENTS, N. AMERICA	C. MURPHY/Y. LECLERC	
172	SMITHSONIAN INSTITUTE	C. MURPHY	
174	DR. J. NAPIER	E. DAHINDEN	
174	BOOK COVER - BIGFOOT, NAPIER	J. NAPIER (EST.)	
175	SKAMANIA ORDINANCE NO. 69-01	PUBLIC DOMAIN	
175	BILL CLOSNER HOLDING A CAST	J. GREEN	
176	BILL CLOSNER & JACK WRIGHT	J. GREEN	
176	AFFIDAVIT OF PUBLICATION	SKAMANIA COUNTY PNR.	
177	SKAMANIA ORDINANCE NO. 1984-2	PUBLIC DOMAIN	
178	GIGANTOPETHICUS & DR. G. KRANTZ	MRS. G. KRANTZ	
178	HUMAN, GORILLA AND GIGANTOPETHICUS SKULL	C. MURPHY	
178	GIGANTOPETHICUS JAW BONES - LEFT PHOTO	MRS. G. KRANTZ	
178	GIGANTOPETHICUS JAW BONES - RIGHT PHOTO	MRS. G. KRANTZ	
179	J. GREEN, D. MURPHY AND GIGANTO SKULL	C. MURPHY	
179	DR. G. KRANTZ AND GIGANTOPETHICUS SKULL	C. MURPHY	
179	GIGANTOPETHICUS SKULL, SEMI-PROFILE	C. MURPHY	
179	GIGANTOPETHICUS SKULL, FULL FACE	C. MURPHY	
179	GIGANTOPETHICUS SKULL FULL PROFILE	C. MURPHY	
180	BIGFOOT XING SIGN	C. MURPHY/J. COOK	
181	R. TITMUS PORTRAIT (UPPER)	R. TITMUS ESTATE	
181	R. TITMUS SEATED AT TABLE	J. GREEN	
182	R. TITMUS "TAKING A BREAK"	J. GREEN	
182	R. TITMUS WITH Dr. G. KRANTZ	J. GREEN	
182	WILLOW CREEK - CHINA FLAT MUSEUM	C. MURPHY	
183	J. GREEN PORTRAIT	J. GREEN	
184	BOOK COVER - SASQ. THE APES AMOUNG US	J. GREEN	
184	BOOK COVER - YEAR OF THE SASQUATCH	J. GREEN	
184	BOOK COVER - THE SASQUATCH FILE	J. GREEN	
184	BOOK COVER - ON THE TRACK OF THE SASQ. BK 1.	J. GREEN	
184	BOOK COVER - ENCOUNTERS WITH BIGFOOT	J. GREEN	
184	BOOK COVER - ON THE TRACK OF THE SASQ.	J. GREEN	
184	BOOK COVER - THE BEST OF SASQ. BIGFOOT	J. GREEN	
185	LYNN MARANDA, C. MURPHY, J. GREEN	C. MURPHY	
185	GREEN AT ABOUT 30-YEARS OLD	J. GREEN	
185	J. GREEN AND DOG (WHITE LADY)	J. GREEN	
185	J. GREEN WITH COMPUTER REPORTS	J. GREEN	
186	DR. G. KRANTZ, I. BOURTSEV AND J. GREEN	J. GREEN	

Page	Title	Credit
186	MELDRUM, STEENBURG, BINDERNAGEL, GREEN	J. GREEN
186	PACIFIC NORTHWEST EXPEDITION GROUP	J. GREEN
186	J. GREEN AND G. HASS	J. GREEN
186	J. GREEN IN HIS OFFICE	C. MURPHY
187	GROUP PHOTO: ON THE ROAD TO WILLOW CREEK	C. MURPHY
188	DR. G. KRANTZ PORTRAIT	HANCOCK HOUSE PHOTO
188	BOOK COVER - BIG FOOTPRINTS	MRS. G. KRANTZ
188	BOOK COVER - BIGFOOT/ SASQUATCH EVIDENCE	MRS. G. KRANTZ
189	DR. G. KRANTZ IN HIS LABORATORY	MRS. G. KRANTZ
189	GROUP PHOTO, MOSCOW	J. GREEN
189	DR. G. KRANTZ AT SYMPOSIUM	C. MURPHY
189	DRS. BINDERNAGEL & KRANTZ	HANCOCK HOUSE PHOTO
190	R. DAHINDIN PORTRAT	C. MURPHY
190	BOOK COVERS (3)	E. DAHINDEN
191	R. DAHINDEN MEASURING TRACKS	E. DAHINDEN
191	R. DAHINDEN WITH R. PATTERSON AT HIS HOME	J. GREEN
191	R. DAHINDEN, R. PATTERSON AND VW VAN	J. GREEN
191	R. DAHINDEN AND CLAYTON MACK	J. GREEN
192	R. DAHINDEN IN LONDON	E. DAHINDEN
192	R. DAHINDEN AND IVAN MARX	J. GREEN
192	R. DAHINDEN AT GARIBALDI PARK	E. DAHINDEN
192	M. DAHINDEN, E. DAHINDEN AND SASQ. CARVING	E. DAHINDEN
192	R. DAHINDEN AND IVAN SANDERSON	E. DAHINDEN
193	D. MURPHY, R. DAHINDEN AND C. MURPHY	C. MURPHY
193	ZENITH SASQUATCH NIGHT AD	E. DAHINDEN/C. MURPHY
193	R. DAHINDEN "IN JOVIAL MOOD"	C. MURPHY
193	R. DAHINDEN AT HARRISON LAKE	C. MURPHY
194	LAND OF THE SASQUATCH PAMPHLET	R. DAHINDEN
194	R. DAHINDEN AND BARBARA WASSON BUTLER	C. MURPHY
195	GROUP PHOTOGRAPH "SASQUATCH DAZE"	C. MURPHY
195	R. DAHINDEN KOKANEE PLACARD	C. MURPHY
195	DAHINDEN, "BREW" AND SASQUATCH PICTURE	C. MURPHY
196	FOOTPRINT IN SAND FOR CASTING	C. MURPHY
196	R. DAHINDEN MAKING CASTS (1ST PHOTO)	C. MURPHY
196	R. DAHINDEN MAKING CASTS (2ND PHOTO)	C. MURPHY
196	R. DAHINDEN MAKING CASTS (3RD PHOTO)	C. MURPHY
196	R. DAHINDEN MAKING CAST (4TH PHOTO)	C. MURPHY
196	R. DAHINDEN SIGNING CASTS	C. MURPHY
196	C. MURPHY & R. DAHINDEN	C. MURPHY
197	CELEBRATION OF LIFE GATHERING	C. MURPHY
197	MEMORIAL CARD	E. DAHINDEN
197	R. DAHINDEN'S STATEMENT	C. MURPHY
198	T. STEENBURG WITH VEHICLE	C. MURPHY
198	FOOTPRINT WITH TAPE	T. STEENBURG
198	MAP - CHILLIWACK AREA	C. MURPHY
199	T. STEENBURG HOLDING CAST	C. MURPHY
199	GROUP PHOTO - WILLOW CREEK JOURNEY	C. MURPHY
199	BOOK COVER - SASQUATCH IN ALBERTA	T. STEENBURG
199	BOOK COVER - SASQUATCH/ BIGFOOT	T. STEENBURG
199	BOOK COVER - IN SEARCH OF GIANTS	T. STEENBURG
199	CAMPSITE	C. MURPHY
199	CHILLIWACK RIVER	C. MURPHY
199	T. STEENBURG "IN THE PATH OF THE FOOTPRINTS"	C. MURPHY
200	D. PEREZ PORTRAIT	D. PEREZ
200	BOOK COVER - BIGFOOTIMES	D. PEREZ
200	NEWSLETTER - BIGFOOT TIMES	D. PEREZ
201	D. PEREZ AND COMPANY INSPECTING FILM SITE	C. MURPHY
201	GROUP PHOTO - PATTERSON/ GIMLIN FILM SITE	D. PEREZ
202	R. NOLL PORTRAIT	C. MURPHY
202	J. GOODALL AND R. NOLL	R. NOLL
203	R. NOLL PHOTOGRAPHING FOOTPRINTS	R. NOLL
203	R. DAHINDEN AND R. NOLL	R. NOLL
203	R. NOLL "IN THE FIELD"	R. NOLL
204	J.R ALLEY PORTRAIT	J.R. ALLEY
205	BOOK COVER - RAINCOAST SASQUATCH	J.R. ALLEY
205	SASQUATCH DRAWING - J. R. ALLEY	J.R. ALLEY
206	R. CROWE PORTRAIT	C. MURPHY
206	THE TRACK RECORD - NEWSLETTER	R. CROWE
207	R. CROWE AND C. MURPHY	C. MURPHY
207	TWISTED TREE BRANCHES	C. MURPHY
207	FAKE CASTS	C. MURPHY
207	R. CROWE'S COLLECTION	C. MURPHY
207	GROUP PHOTO - CARSON	C. MURPHY
208	M. MONEYMAKER PORTRAIT	M. MONEYMAKER
208	BFRO LOGO	BFRO
209	B. SHORT PORTRAIT	B. SHORT
210	JOHN CHAMBERS	B. SHORT
211	P. SMITH PORTRAIT	C. MURPHY
211	POSTER - SASQUATCH FAMILY	P. SMITH
212	POSTER - SASQUATCH FORAGING FOR FOOD	P. SMITH
213	POSTER - SASQUATCH RUNNING	P. SMITH
214	RUSSIAN SNOWMAN DRAWINGS	D. BAYANOV
214	RUSSIAN SNOWMAN FOOTPRINT - TIEN SHAN	D. BAYANOV
214	RUSSIAN SNOWMAN FOOTPRINT - DOLINA N.V.	I. BOURTSEV
215	D. BAYANOV IN HILLS OF KABARDA	D. BAYANOV
215	LT. COL. V. S. KARAPETIAN	V. KARAPETIAN
215	BOOK COVER - IN THE FOOTSTEPS OF THE R.S.	D. BAYANOV
215	KARAPETIAN EXTENDS HIS HAND	L. BOURTSEVA
216	KARAPETIAN HOMINID	UNKNOWN ARTIST
216	BOOK COVER - AMERICA'S BIGFOOT	D. BAYANOV
216	BOOK COVER - BIGFOOT: TO KILL OR TO FILM	L. BOURTSEVA
217	BOURTSEV HOLDING KHWIT'S SKULL	I BOURTSEV
217	KHWIT'S SKULL	I. BOURTSEV
217	KHWIT PHOTO	I. BOURTSEV
217	DR. G. KRANTZ, SKULL & I.BOURTSEV	MRS. G. KRANTZ
217	KHWIT PHOTO	I. BOURTSEV
219	HORSE AND FARMER - BRAIDED MANES - TOP	D. BAYANOV
219	HORSE AND FARMER - BRAIDED MANES - LOWER	D. BAYANOV
220	I. BOURTSEV IN A DIG	D. BAYANOV
221	RUSSIAN CHILDREN - RURAL AREA	D. BAYANOV
222	NINA GRINYOVA TALKING TO WITNESSS	D. BAYANOV
223	D. BAYANOV, D. DONSKOY, I. BOURSEV	I. BOURTSEV
223	I. BOURTSEV & D. BAYANOV	D. PEREZ
223	D. BAYANOV WITH "PATTY"	E. DAHINDEN/C.MURPHY
224	SASQUATCH SCULPTURE IN SETTING	I. BOURTSEV
224	GROUP PHOTO - FOUNDERS	D. BAYANOV
224	POSSIBLE SNOWMAN SHELTER	I. BOURTSEV
224	I. BOURTSEV COMPARING HIS FOOT TO CAST	I. BOURTSEV
224	I. BOURTSEV WITH KIROV REGION HUNTSMEN	I. BOURTSEV
225	YETI FOOTPRINT CAST	C. MURPHY
226	POSTAGE STAMPS - YETI, BHUTAN	NOT KNOWN
226	POSTAGE STAMP SHEET - MALDIVES ISLANDS	MALDIVES ISLANDS POST
227	YETI PAINTING	R. BATEMAN
229	R. GIMLIN & C. MURPHY	C. MURPHY
000	BACK COVER - C. MURPHY	C. MURPHY
000	BACK COVER - J. GREEN	C. MURPHY
000	BACK COVER - T. STEENBURG	C. MURPHY

NOTE 1:
COPY OF CAST(S) IN COLLECTION OF J. GREEN, B.C.

NOTE 2:
COPY OF CAST IN COLLECTION OF J. BINDERNAGEL, B.C.

NOTE 3:
COPY OF CAST IN COLLECTION OF T. STEENBURG, B.C.

NOTE 4:
(L) COPY OF CAST IN COLLECTION OF J. GREEN; B.C., (R) E. DAHINDEN, B.C.

NOTE 5:
CASTS MADE BY BONECLONES, CA

NOTE 6.
COPY OF CAST IN COLLECTION OF R. CROWE, OR.

GENERAL INDEX

A

Abbott Hill, Washington, 105, 114
Abbott, Don, 41, 49, 111, 132
Adams County, Ohio, 144
Agassiz, 32
Aggassiz-Harrison Advance (newspaper), 183, 185
Akron, Ohio, 159
Alaska Division of Forestry, 158
Alaska Fish and Game (Dept.), 158
Alley, Robert J., 156, 204
American Anthropological Research Foundation Inc., 14
American National Enterprises, 86
American Yeti Expeditions, 14
Anahim Lake, B.C., 191
Anderson, Alexander Caufield, 32
Anderson, California, 181
Arcata, California, 43
Argosy magazine, 50
Assu, Don, Chief, 22
August, Adaline, 32
Austin, Ned, 27

B

Baird, D., 141
Baku, Azerbaijan, 218
Barnum and Bailey, 27
Barnum, P.T., 27
Bateman, Robert, 97, 100, 227
Bauman, 25
Bayanov, Dmitri, 72, 81, 83, 99, 165, 180, 187, 189, 199, 201, 215, 216, 219, 220, 221, 222, 223, 224
Baylor Medical Center, Texas, 209
Beacon Rock State Park, Washington, 176
Bear Creek, Washington, 176
Beck, Fred, 28, 29, 183
Beebe, Frank, 49, 161
Beijing, China, 204
Bella Coola, B.C., 13, 39
Bellingham, Washington, 35, 38
Benassi, Mario, 91
Bering Strait, Alaska, 16, 49
Berry, Al, 160
Berryman, Bruce, 115
BFRO, 145, 171, 203, 208
Bigfoot Encounters (website), 210
Bigfoot Field Researchers Organization, 145, 171, 203, 208
Bigfoot Research Project, The, 80
Bigfoot Times, 200
Bindernagel, John, Dr. 105, 145, 162, 163, 180, 186, 189, 201
Birnam, Penny, 71
Bitterroot Mountains, Idaho/Montana, 25
Blanchard, Idaho, 138
Blue Creek Mountain, 41, 105, 110, 111, 121, 132, 137, 191
Blue Mountains, Washington, 129, 137, 138, 142, 205
Bluff Creek, California 34, 37, 41, 43, 44, 45, 47, 50, 58, 64, 72, 76, 84, 92, 102, 103, 105, 106, 107, 114, 115, 116, 119, 120, 173, 181, 182, 191, 196, 201, 202, 228
BoneClones (company), 179
Bord, Janet, 180

Bossburg, Washington, 107, 117, 134
Bourtsev, Igor, 57, 72, 81, 96, 180, 186, 189, 201, 216, 217, 223, 224
Bourtseva, Alexandra, 189, 218
British Columbia Expedition, 38, 39
British Columbia Provincial Museum, 41, 49, 132, 161
Brudevoid, Joe, 113
Buck' was, 17, 18
Buncombe Hollow, Washington, 119
Burchak-Abramovich, Nikolai, Prof., 219
Burney, California, 192
Burns, John W., 31, 38
Butler, Barbara Wasson, 194, 195
Buttle Lake, B.C., 204
Byrne, Peter, 85, 124

C

Cairo, Leonard, 175
Cameron, Constance, 180
Campbell River, B.C., 10, 22
Canada Post Corporation, 24, 101
Canada's Legendary Creatures (postage stamps), 101
CanWest film facility, 85
Casterton, Mr., 26
Cates, Charles, 38
Central Institute of Physical Culture, Moscow (Russia), 74
Chambers, John, 84, 209, 210
Chapman, George, 35
Chapman, Jeannie, 35
Chehalis River, B.C., 34
Chehalis, 17, 31, 33, 38
Chehalis, Phillip, 33
Chilcutt, Jimmy, 128, 140, 141, 180, 188
Chilliwack River, B.C., 106, 198, 199
Chilliwack, B.C., 182
Chinook Press, 88
Clappison, George, 154, 158, 159, 180
Closner, Bill (Sheriff), 175, 176
Cochiti, 14
Coleman, Loren, 180
Columbia University Graduate School of Journalism, 185
Conroe Police Department, Conroe, Texas, 140
Cook, Joedy, 106, 154, 159, 180
Cottonwood, California, 114
Cowan McTagert, Ian, Dr., 49
Cox, Don, 176
Craft, Roy, 175
Craig, Alaska, 157
Craig, R. J., 27
Cranbrook, B.C., 85
Crew, Jerry, 37, 41, 105, 106, 115, 116, 181
Crook, Cliff, 87
Crowe, Ray, 206, 207
Cryptozoology (Journal), 124, 141
Cryptosphere, 223
Custerton, Mr., 27

D

Dahinden, Erik, 68, 192
Dahinden, Martin, 192
Dahinden, René, 30, 34, 35, 38, 39, 41, 48, 49, 58, 59, 68, 69, 73, 74, 76, 86, 101, 107, 110, 117, 181, 183, 186, 190, 191, 192, 194, 195, 200, 201, 203, 204, 205, 219, 224
Daily Mail (newspaper), 190
Dallas, Texas, 209
Darwin Museum, Moscow, Russia, 215, 218
Davenport, Everett, 29
Davis, Carol, 117
Davis, Jack, 194
Davis, M.K., 180
Davis, Norm, 117
De Atley, Al, 45, 84
Delaware (First Nations People), 18
Delox Marsh, Wisconsin, 121
Discovery Channel, 90
Dolina Narzanov Valley, North Caucasus, 215
Donskoy, Dmitri d., Dr., 74, 83, 189, 223
Drever, Lee, 185
D'sonoqua, 14, 17, 22
Dunn, Joe, 35, 38

E

Early, George, 180
Elk Wallow, Walla Walla, Washington, 107
Ellensburg, Washington, 112
Elma Gate, Washington, 114
Enders, Terry, 159
Estacada, Oregon, 121, 161
Eurasia, 16
Eureka, California, 43, 45, 114, 120

F

Fahrenbach, W. Henner, Dr. 81, 87, 124, 128, 143, 155, 180
FBI, 113, 152, 153, 154, 155
Federal Bureau of Investigation (FBI), 113, 152, 153, 154, 155
Fifth European Symposium of Cryptozoology, 210
Fish, Leroy, Dr., 145
Flint River, Georgia, 140
Flying Eagle, Chief, 32
Fort Braggs, California, 144
Fort Langley, B.C., 151
Frakes, Jonathan, 160
Franzoni, Henry, 180, 210
Fraser Valley, B.C., 182
Freedom of Information – Privacy Act, 154
Freeman, Paul, 107, 109, 128, 142, 143, 144, 185, 205
Freemont, Wisconsin, 121

G

Gagit, 18
Garibaldi Park, B.C., 192
Gashpeta, 14
Gear, R. J., 79
George Inlet, Alaska, 205
Gerasimov, Mikhail, 220
Gibbs, S. H., 12
Gigantopithecus blackie, 178, 188, 204
Gilford Pichot National Forest, Washington, 145, 150
Gilyuk (First Nations name for sasquatch), 98
Gimlin, Robert, 40, 41, 42, 43, 44, 45, 47, 49, 50, 63, 64, 73, 84, 86, 87, 89, 92, 128, 131, 180, 183, 187, 199, 201, 228, 229, 230
Glickman, J. (Jeff), 80, 81, 83

Goodall, Jane, Dr., 202
Gouin, Mr., 27
Graves, Pat, 47, 106
Grays Harbor County, Washington, 114, 133
Green River Community College, 202
Green, Beverly, Dr., 49
Green, Jim, 161
Green, John, 12, 16, 38, 39, 41, 49, 59, 64, 65, 79, 81, 86, 86, 89, 105, 109, 111, 114, 116, 124, 128, 137, 141, 145, 167, 171, 173, 179, 181, 183, 184, 185, 186, 187, 189, 191, 194, 199, 201
Greene County, Arkansas, 25
Grieve, Donald W., Dr., 57, 76, 79, 81, 83
Grinyova, Nina, 222
Guiguet, Charles, 49

H

Hairy giants, 31
Hairy man, 19, 20, 21
Hajicek, Doug, 90, 92, 93
Hamilton, Mr., 24
Hancock, Dave, 49, 64, 182
Hannington, Dr., 27
Harrison Hot Springs, B.C., 31, 124, 128, 181, 182, 183, 186
Harrison Lake, B.C., 32, 182, 193, 203
Harrison River, B.C., 31, 38
Harvey, Stephen, 189
Hazelton, B.C., 181, 182
Heinselman, Craig, 180
Heironimus, Robert, 89
Herald of Hominology, The, 223
Hereford, Jo Ann, 180
Heryford, A. Dennis, 105, 114, 133
Hewkin, J.A., 124
Hillsboro, Oregon, 206
Hocheu, 20
Hollywood, California, 84
Hope, B.C., 198
Horseman, M., 141
Hubbard, Ohio, 116
Hudson's Bay Co., 32
Humboldt County, California, 43
Humboldt Times (newspaper), 37
Hunter, Don, 74, 190
Huseinov, Shahbala, 218
Hyampom, California, 106, 115, 138

I

Idaho State University, 129
International Bigfoot Society, 206

J

Jacko, 26, 27
Jasper, Albert, 24
Jeftichew, Fedor, 27
Johnson, Kirk, 39, 186
Jo-Jo the Dog-Faced Boy, 27

K

Kalika, Pinya, 218
Kane, Paul, 24
Karapetian, Lt., Col., MD, 215, 216
Karbarda, North Caucasus, 215
Keating, Don, 180
Kemball, Harry, 84, 85, 86
Kenmore, Ohio, 159
Ketchikan, Alaska, 205
Khwit, 216, 217, 220, 221
Kirlin, R. Lynn, Dr., 160

KIT radio (station), 88
Kitimat, B.C., 181, 182
Kiviat, Robert, 89
Klawock Lake, Alaska, 156
Klawock, Alaska, 157
Klemtu, B.C., 39, 181
Koffmann, Marie-Jeanne, 189, 217, 218, 224
Korff, Karl, 89
Koval, Gleb, 222, 223
Krantz, Grover, S. Dr., 12, 49, 69, 81, 82, 83, 107, 113, 117, 119, 120, 129, 138, 141, 145, 178, 179, 182, 186, 188, 189, 196, 201, 205, 214, 217
Kwakiutl, 14, 18, 22

L

Labatt Brewing Company, 195
Laird Meadow Road/area, Bluff Creek, California, 47, 106, 144
Lake Chopaka, Washington, 91
Lalooska, 22
Landis, John, 209
Laverty, Robert Lyle, 48, 131
Lebedeva, Galina, 220
Leclerc, Yvon, 11, 66, 67, 71, 81, 98, 100, 108, 143, 164, 165, 166, 171
Legend of Boggy Creek, The (documentary), 200
Leiby, Richard, 89
Leick, Robert K., 175
Lewis River, Washington, 28
Lillooet, B.C., 12
Lindley, Caroline Sue, 144
Logan, Nancy, 160
London, England, 27
Long, Greg, 89
Long, Serephine, 33
Louse Camp, California, 41, 186
Lummi peninsula, 113
Lund, Larry, 180, 195
Lundy, Conrad, 175
Lytton, B.C., 26

M

Mack, Clayton, 39, 191
Macleod, James, Dr., 138
Maidar, (prime minister – Mongolia), 221
Major, Mr., 27
Makarov, Vadim, 189, 222
Malone, Tom (attorney), 89
Maranda, Lynn, 185
Markotic, Vladimir, Professor, 204
Marsh, Q.C., 10
Marx, Ivan, 109, 117, 185, 192
Mashkovtsev, Alexander, Dr., 216, 224
Mayak datat, 22
McCall, Rob, Dr., 226
McClarin, Jim, 180, 192
McCormick, Jim, 89
McCoy, Syl, 112, 115
McEntire, Mac, 87, 88
McFarland, Lee, 180
McKelvie, Bruce, 38
McLean's (magazine), 194
Meldrum, Jeffrey, Dr., 104, 129, 139, 141, 145, 151, 173, 180, 186, 201
Memorial Day footage, 95
Memphis, Tennessee, 25
Merwin Dam Reservoir, 119

Mi!yak datr!atr!, 22
Mica Mountain, B.C., 36
Miles, John, 195
Milner, Robert, 195
Minneapolis, Minnesota, 90
Mission, B.C., 198, 205
Moneymaker, Matt, 201, 208
Montgomery, Bob, 180
Moorman, Eliza, 119
Morehead, Ron, 160
Morgan, Robert W., 14, 15, 20, 119, 120, 175, 180
Morris Mountain, B.C., 32, 33
Moscow, Russia. 73, 217
Moskowitz, Kathy, 19, 23
Mount St. Helens, Washington, 24, 28, 41, 116
Mountain giants, 30
Muench, Eric, 156
Murphy, Christopher, 87, 187, 196, 207
Murphy, Christopher, Jr., 193
Murphy, Dan, 34, 101, 128, 179, 193, 195
Mysterious Encounters (TV series), 90

N

Napier, J. R. (Dr.), 124, 172, 174
NASI, 80, 81, 83, 210
Neiss, Tod, 180
Newberry, Sally, 180
Nimpkish, 17
Niska, 17
Noll, Richard, 105, 109, 145, 151, 202
Nooksack River, 113
Nordegg, Alberta, 204
North American Science Institute (NASI), 80, 81, 83, 210
North Idaho College, 138
North Vancouver, B.C., 85
Northwest Territories, Canada, 90
Norwalk, California, 200
Notice Creek, California, 41, 186
Notice Creek, California, 41, 44
Nunez, James, 205

O

Oching'-i-ta, 19
Offield Mountain, California, 118
Ohio Bigfoot Nest, 158
OLN, 90
Okladnikov, Alexey, Dr., 221
Onderdonk, Mr., 26
Onion Mountain, California, 119, 144
Orchard, Vance, 205
Oregon Primate Research Center, 124
Orleans, California, 118
Ostman, Albert, 30, 151, 183
Outdoor Live Network, 90
Oxford Institute of Molecular Medicine, 226

P

Pacific Northwest Expedition, 38, 39, 118, 183
Painted Rock, California, 13, 19, 20, 21
Parmir Mountains (Russia), 153
Pate, Lori, 95
Pate, Owen, 95
Patrick, Ed, 39, 186
Patterson, Patricia, 41, 50, 89
Patterson, Roger, 40, 41, 42, 43, 44, 45, 46, 47, 48, 49, 50, 55, 56, 57, 63, 66, 73, 76, 84, 85, 86, 87, 89, 92, 102, 106, 107, 131,

172, 183, 189, 191, 194, 196, 228, 229
Paul, Alexander, Chief, 34
Pei, Professor, 204
Perez, Daniel, 58, 59, 195, 200, 201, 223
Planet of the Apes (movie), 84, 209
Poinsett County, Arkansas, 25
Point, Ambrose, 17
Point, William, 32
Porshnev, Boris, Dr., 216, 218, 219, 220, 221, 224
Porter Creek, Washington, 114
Porterville, California, 19
Powder Mountain, B.C., 121
Powell, Thom, 145, 180
Prachtengertz, Michail, 189
Princeton Museum, B.C., 12
Prior, Derek, 91, 95
Pullman, Washington, 205

Q
Qualac, 33
Quast, Michael, 180

R
Randles, Derek, 145
Redding, California, 38
Reiter, William, 195
Relict Hominoid Research Seminar (also Smolin Hominology Seminar), Moscow, Russia, 215, 218, 223
Rense, Jeff, 89
Rhodes, Joe, 117
Rice, Steve, Dr., 154
Rienke, Clyde, 86
Rinchen (academician), 221
Rio Grand River, New Mexico, 14
Roe, William, 36
Romney, Jerry, 86
Roosevelt, Theodore, 25
Royal Free Hospital School of Medicine (England), 76
Rubec, Peter, Dr. 180
Ruby Creek, B.C., 35, 183
Rugg, Michael, 180
Russian Cryptozoology Association, 223

S
Salt Fork State Park, New Mexico, 18
San Antonio, Texas, 38, 39
San Francisco, 37
San Joaquin Valley, California, 19
San Jose News (newspaper), 120
Sanderson, Ivan T., 38, 39, 40, 50, 192, 194, 204
Sarmakovo, Russia, 217
Sarmiento, Esteban, Dr., 145, 151,
Saskahana, 33
Sasquatch: Legend Meets Science (documentary), 90, 95
Satsop River, Washington, 133
Sawvel, John, 159
Schaffner, Ron, 180
Schaller, George, Dr., 145, 202
Scott Stamp Monthly (magazine), 101
Seattle, Washington, 211
Sequoia National Forest, California, 21
Sergeev, Valery, 224
Seskehavis, 31
Shackley, Myra, 221
Shawnee State Park, Ohio, 106

Shipton, Eric, 226
Shoonlshoonootr!, 20
Short, Bobbie, 84, 209
Sierra Mountains, California, 160
Simon Fraser University, 205
Skamania County Ordinance, 175, 177
Skamania County, Washington, 119, 175, 176, 177
Skeena River, B.C., 106, 118, 182
Skoocooms, 24
Skookum cast, 145, 203
Skookum Meadows, Washington, 145
Slick Airways, 39
Slick, Tom, 38, 39, 181, 186
Slick, Tom, Sr. 39
Smith, David, 202
Smith, Marion, 29
Smith, Paul, 123, 211, 212,213
Smithsonian Institution, 73, 172, 173, 174, 189
Smolin, Pyotr, 224
Smolin Hominology Seminar (also Relict Hominoid Research Seminar), Moscow,
Sykes, Bryan, Professor, 226
Russia, 218, 223
Sonora, California, 21
Southwest Foundation, 39
Sprague, Roderick, 10
Spuzzum Flats, B.C., 27
St. Francis County, Arkansas, 25
Standard-Weekend Magazine, The, 194
Stanislaus National Forest, California, 21
Steenburg, Thomas, 106, 109, 186, 187, 198, 199, 205
Steindorf, Reuben, 90, 92
Strathcona Provincial Park, B.C., 105, 204
Summerlin Wes, 205
Sunsunut, 22
Susemiehl, John, 117
Swanson, Bob, 88
Swindle Island, B.C., 39
Swindler, Daris, Dr., 145, 151

T
Table Springs, Washington, 138
Tampico, Washington, 191
Tatsl, Igor, 221, 222
Tench, Charles V., 31
Tero, Bill, 175
Terrace, British Columbia, 118
Tete Jaune Cache, B.C., 36
Thomas, Glen, 161
Thompson, David, 24
Thompson, Warren, 195
Thorington, Jr., Dr., 73
Tien Shan, Russia, 214
Tilbury, George, 27
Times, The (British newspaper), 226
Times-Standard, The (newspaper), 43, 45
Titmus, Bob, 38, 39, 42, 49, 103, 106, 107, 112, 115, 116, 118, 131, 138, 144, 181, 186, 195, 196, 230
To the Ends of the Earth (documentary series), 226
Toba Inlet, B.C., 30
Tombe, Brad, 34
Tout, Hill, 32
Track Record, The, 206

Trachtengerts, Michael, 222
Travers, Pete, 16, 145, 148
Tripp, R. Maurice, Dr., 120
True (magazine), 40, 194
Trumbull County, Ohio, 116
Tsimshian, 17
Tsungani, 22
Tule River, California, 13, 19, 21
Twan, Wanja, 197
Twisp, Washington, 202

U
Uchiyingetau, 19
Umatilla National Forest, Washington, 129
United States Army Corps of Engineers, 152
University of British Columbia, 49, 172, 185, 229
University of Calgary, 204, 205
University of Idaho, 10
University of London (England), 174
University of Washington, 151
University of Wyoming, 160

V
Vancouver Province (newspaper), 37
Vancouver, B.C., 31, 37
Vander White, 28
Victor, Charley, 33
Victoria, 26
Vision Realm (company), 90

W
Walla Walla River, Washington, 138
Walla Walla, Washington, 107, 129
Wallace, Ray, 84, 109, 184, 207
Walsh, Gerri, 39, 186
Ward, Michael, 226
Washington Environmental Atlas, 152, 154
Washington State University, 82, 129, 188, 196
Washougal, Washington, 176
Wasson, David, 87
West Valley, Yakima County, 88
Western Bigfoot Society, 206
White Wolf Entertainment Inc., 90
White, Thos., 27
Wild Entertainment Inc., 100
Wilderness Productions Inc., 100
Williams, Peter, 33
Willich, Wilbur, 190
Willow Creek – China Flat Museum, 182, 192
Willow Creek, California, 19, 38, 44, 64, 128, 129, 194
Winnipeg, Manitoba, 204
Woodard, Harry M. (attorney), 88, 89
Woodland, Washington, 119
Wooly-Boger, 20
Workman's Bar, Washington, 114
Wright, Jack (Deputy), 176

X
X Chronicles, 84, 85, 86

Y
Yakima City, Washington, 87, 88, 89
Yakima County, Washington, 40, 43, 45, 50
Yale, B.C., 26, 27
Yokuts, 19, 20, 21

Z
Zana, 215, 216, 217, 219, 220, 221
Zenith Masonic Hall, 193
Zilla, Washington, 88

Sasquatch: The Apes Among Us
John Green
0-88839-018-1
5½ x 8½, sc, 492 pages

Bigfoot Sasquatch Evidence
Dr. Grover S. Krantz
0-88839-447-0
5½ x 8½, sc, 348 pages

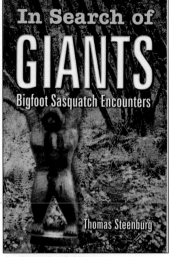
In Search of Giants
Thomas Steenburg
0-88839-446-2
5½ x 8½, sc, 256 pages

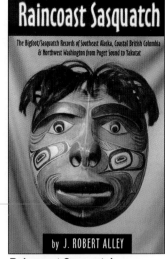
Raincoast Sasquatch
J. Robert Alley
0-88839-508-6
5½ x 8½, sc, 360 pages

The Locals
Thom Powell
0-88839-552-3
5½ x 8½, sc, 271 pages

Strange Northwest
Chris Bader
0-88839-359-8
5½ x 8½, sc, 144 pages

The Best of Sasquatch Bigfoot
John Green
0-88839-546-9
8½ x 11, sc, 144 pages

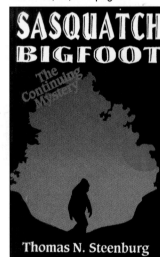
Sasquatch Bigfoot
Thomas N. Steenburg
0-88839-312-1
5½ x 8½, sc, 126 pages

The Bigfoot Film Controversy
Roger Patterson and Chris Murphy
0-88839-581-7
5½ x 8½, sc, 240 pages

In the Footsteps of the Russian Snowman
Dmitri Bayanov
5-900229-18-1
5½ x 8¼, sc, 239 pages

America's Bigfoot: Fact, Not Fiction
Dmitri Bayanov
5-900229-22-X
5½ x 8½, sc, 224 pages

All titles available at:
HANCOCK HOUSE PUBLISHERS
1431 Harrison Avenue, Blaine,
WA 98230-5005, USA
(604) 538-1114 Fax (604) 538-2262
www.hancockhouse.com
sales@hancockhouse.com